The Instrument Pilot's Library
• Volume One •

Sweet Talking the System

by

The Editors of *IFR* and *IFR Refresher*

Belvoir Publications, Inc.
Greenwich, Connecticut

ISBN: 1-879620-14-6

Printed and bound in the United States of America by Arcata Graphics, Fairfield, Pennsylvania.

Contents

Preface

There comes a time in every flight when you have to face the reality of what you've begun and actually taxi onto the runway and depart. Until then, your airplane is nothing more than a vehicle and you, a driver.

Making your way through clearance delivery is child's play. Ground control offers little more challenge, with a two-dimensional clearance to the active: "Taxi via inner, outer, T-1, cargo, give way to the Northwest 'Nine', hold short of runway 23."

Yeah, yeah, yeah, you read back and crack the throttle. Your Bonanza sounds more like an old Harley than a six-figure flying machine. Loping along, you run through the switches, adjust your seat once more and tap your knuckle on the CHT. It's something you do; it's something all pilots do, and you desperately want to do what all pilots are supposed to do.

But as you approach the runway a transition begins. You're about to become unglued from the earth and step into the lonely third dimension of IFR flight. It's here that your only connection with the earth is your voice, and what you say and how determines your status in the ATC system.

Forget that your airplane has a loran with a database that includes the names and addresses of every pilot in the free world. Once you taxi onto the active, your entire existence falls into the hands of a middle-aged air traffic controller who drives a '78 Datsun wagon with no muffler and thinks of you as one more piece of information to be pushed across a radar display. It becomes a case of your voice and his voice getting you to your destination.

So now you're on the runup pad, ready to go:

Tower: "Bonanza Two Eight Delta, fly runway heading; runway 30R, cleared for takeoff."

You: "Runway heading, and cleared to go."

You heard what tower said, but what about that SID in your clearance? It says nothing about runway heading. No problem. Change of plan. Stay flexible. Disregard the SID for now, reset the bug and fly runway heading.

Throttle open, a quick glance at the gauges—all green. There's your speed, rotate, climb, gear up. Tower barely notes your passing into the cloud base. Eyes on the panel, you fly runway heading and climb to assigned altitude. Everything looks good, so it's time to switch frequencies. You punch the flip-flop.

You: "Departure, Bonanza Six Two Eight Delta with you leaving one point four for three, flying runway heading."

Departure Control (slightly edgy): "Two Eight Delta, turn left—now—heading two-seven-zero, climb and maintain three thousand. Start your turn now!"

You (confused, but complying): "Ah, left to two-seventy and up to three, ah, Two-Eight Delta."

Departure (belatedly): "Bonanza Two-Eight Delta, radar contact." Then, this: "And Two Eight Delta, in the future do not change frequencies until tower tells you to."

You (embarrassed but uncertain why): "Ah, roger, departure, ah...." (What's he talking about?)

Departure (sounding a lot like Mr. Mathews, your 9th grade biology teacher): "Two-Eight Delta, tower had traffic for you, and needed you on that two-seventy heading before they shipped you to departure." (You try to apologize, but he's still talking.)

"By changing frequencies on your own, he couldn't turn you, and you were eating up a Cessna in the clouds ahead of you."

Your blushing appears on the radar scope as a faint magenta ring around your target, plus you've wandered away from your assigned heading by at least 40 degrees, and you're about to blow through 3000 feet. (Won't he quit about the frequency business!)

"And, Two-Eight Delta, check your DG, fly heading Twooo-Sevennn-Zeroooo."

That pretty much sets the mood for the flight. Even though you thought you were on top of this ATC stuff, and didn't want to sound like an average pilot, you've managed to get caught in the full glare of an FAA voice. You've ticked-off a controller, and you're certain your registration number will be entered into a secret ATC vendetta file in

Washington, D.C., and from that day forward, wherever you fly, you'll have to plan an extra 20 minutes fuel for punitive vectors.

Keeping your nose clean

Every pilot who's fouled up when it comes to ATC (and if you're one that hasn't—ever—you can apply for sainthood at the nearest FSDO...and send your name to *Ripley's Believe It or Not*) knows what our hapless Bonanza driver is going through.

It's the desire of every IFR pilot to get where he or she is going with a minimum of fuss and bother. Ideally, that involves a painless trip through the air traffic control system and favorable weather. While we can't do anything about the weather, we can help you improve your chances of getting through the system as easily and quickly as possible, while keeping both yourself and the controllers happy.

There's more than one way to do it. You can muddle through in a perfectly legal manner and complete a given flight, but it might entail delays, reroutings, hassles and short tempers. Or, you could cut through the nonsense...providing you know how.

The path of least resistance

In this first volume of the *Instrument Pilot's Library*, our purpose is to give you the tools you need to make the system work for you, instead of allowing it to have its way with you.

The training leading up to your instrument ticket teaches you some of how the system works from the pilot's viewpoint, but the details necessary to take full advantage of it are rarely covered. Often, the only way to find out how best to work your way through the system is through many hours of experience. We intend you help you learn the ins and outs without having to do it the hard way.

That's not to say you'll always win the battle of trying to get direct to your destination, but if you take the proper steps you're far less likely to have to put up with clearances that take you over West Wheretheheckarewe. You'll get where you're going sooner, and with less aggravation for both you and the controller if you know how to get what you want.

This book deals with a variety of topics from the point of view of both the controller and the pilot. They're quite different: it's a common misconception that controllers know what it's like in the cockpit. There's no requirement for a controller to also be a pilot, or to fly at all, for that matter. Many have no better understanding of what it's like to be a pilot than pilots have of what being an air traffic controller entails.

What we'll cover

Sweet Talking the System offers specific, hard advice on how to navigate the air traffic control system, but it also goes beyond, lending insight into how the system works and why things happen the way they do.

In the first section, "Inside ATC," we'll give you a rare look at air traffic control, from how controllers are trained to what it's like to be one, to a close examination of how the system works.

In "Unraveling Clearances," we'll cover the importance of clear, unequivocal clearances and how to deal with unusual instructions.

Next, in "Getting the Route You Want," pilots and controllers give detailed instruction in how to make the most of your time in the cockpit: how to beat the preferred route, the value of VFR flight, and how to get direct routings even if you're not equipped with an RNAV. We'll also discuss many of the factors that determine why controllers do what they do, and the limits of the system.

Finally, in "Communications," we cover the instrument pilot's most important skill, talking on the radio.

Who we are

Each of the books in the *Instrument Pilot's Library* contains first-hand knowledge from the editors of *IFR* and *IFR Refresher*. This highly qualified group includes ATPs, controllers, charter pilots, instructors, and ordinary instrument pilots, giving you the widest variety of view-points possible.

• Section One •

Inside ATC

The Making of an Air Traffic Controller

Much is written about achieving and maintaining an instrument rating. Flight schools compete for the student market, making it appear as though the key to success is a wheelbarrow full of money. But what about the voices at the other end of the radio? How are controllers made?

A few schools, such as the University of North Dakota, do offer ATC curricula but most ATC training is performed by the largest ATC employer—the FAA. True, some non-federal towers train controllers and so does the military but neither guarantees a job as a genuine FAA controller. If you want to be an FAAer, you have to follow the government program, a process that can be long and frustrating.

Like pilot training, ATC training is spelled out in a pile of orders and directives which are constantly revised, updated, re-revised and reviewed. It's difficult to say with any certainty what the FAA's training philosophy is at any given moment. Catch phrases come and go but the process follows a general path that leads, sooner or later, to the FAA Academy in Oklahoma City.

IFR *contributor Paul Berge, a tracon controller in Des Moines, graduated from the Academy in 1979. Here he offers an insider's view of how ordinary citizens become air traffic controllers.*

Making It to the Big Time

The ATC hiring and training process is undergoing a major overhaul. The "good old boy" network, dating back to the round engine and brown shoe days, is going the way of the colored airways. Today, an applicant gets his or her foot in the door through Civil Service, followed by a new, high-tech screening program. Not that this will change ATC much. It'll still be "cleared for takeoff, cleared to land, turn left, turn right." Only the hoops through which the perspective controller must jump will be rearranged.

If you're a U.S. citizen, between 21 and 31, with a minimal criminal record and a clean bladder, you are eligible to apply for ATC. First comes the Civil Service entrance exam, a test much like any other extremely long and pointless broad skills test designed by experts who know nothing about controlling airplanes. You'll be expected to unravel dot patterns and decode number sequences and fill the correct answer on a standard answer sheet with a No. 2 pencil. Anyone born after 1947 should be familiar with the routine.

Many months after your exam, you will receive a card, similar to pilot written exam results. While 70 percent is passing, you will need at least 90 percent to be considered for the "register." The register is a list of who is eligible should an opening arise. Currently, there are more than 5000 names on the register awaiting ATC slots.

Hiring begins at the top and works down. It tends to run in spurts in the FAA. That's not the agency's fault. Political winds dictate budgets and the FAA goes through long dry spells when money is tight, followed by, "Oh, M'Gosh! We need people—fast! Hire!" Then the money flies and the Academy runs double shifts to handle the influx.

There are ways to circumnavigate the process. College students can apply for a "co-op" program. Students take the same Civil Service test and they spend the next two summer semesters working in an ATC facility. They're run through ATC classes and (sometimes) allowed to work traffic. After completing 26 weeks with the FAA and graduating from their college with at least a 2.0 GPA, they're eligible to enter the FAA Academy. If there are any openings.

Before entering the co-op program, the student must pass a Pre-Training Screen (PTS). All prospective controllers must pass the screen. For many, it's their first and last encounter with the FAA.

The screen

When *IFR* contributor Scott Dunham and I went through the screen together in 1979, it consisted of nine weeks of misery in Oklahoma City. We ran through endless non-radar air traffic problems wherein one student would play pilot while the other student played controller. At the end of the nine weeks (the same length, coincidentally, as my Army basic training), the student had either passed or failed. Passing meant a job; failure a hearty handshake. With the advance of the computer age, the screen is now reduced to five days.

Today's would-be controller—not yet hired—arrives at FAA expense in Oklahoma City. Housing is provided in a local hotel. Meals are also provided. Hope you like Spam. On Monday, the candidate is brought to the Academy at the Mike Monroney Aeronautical Center.

The Aeronautical Center is the FAA's Mecca. Everything from your airman medical records to spare VOR radials are located there. FSDO, Airway Facilities, Flight Service, Flight Check and Air Traffic all run schools at the Center. They even have a unit trying to figure out how to get a million pilots to put their pictures on their pilot certificates each year.

Back to the screen. For the first three days, the candidate will sit in front of a computer screen to practice vectoring little targets through gates and into airports. This is not real ATC. It is a test. The computers are voice activated. The candidate puts in a full eight-hour day, of which six hours are spent at the computer.

The video tasks are varied. One requires the operator to unscramble dot patterns while estimating how long a separate dot takes to move, unseen, across the screen. All tasks involve performing several functions at once in a timed environment. The pressure is applied on the first lesson and doesn't ease up until after the final test.

The screen's objective is to get all candidates to the same computer skill level before the actual tests are given. Testing for prospective controllers is more complex than just deciding if a person has the ability to separate airplanes. That would be relatively easy. Run a candidate through a quick ATC class and sit them in a radar simulator and you'd have some idea.

Broader political and legal considerations have to be considered when hiring into a government job. To simply take the "best qualified" person—e.g. experienced pilot or military controller—might continually discriminate against individuals who may have never been exposed to aviation.

By creating a screen in which all candidates are brought to the same level of "experience" prior to testing, the playing field is leveled somewhat. Whether or not you'll get the best qualified controller in the end is someone else's call.

On Thursday, the candidate is tested. The same tasks practiced for the previous three days are performed and a score is earned. Make enough points and you pass. Lose too many points running dots together, and you're on the Friday afternoon Greyhound.

The screening I saw on my visit late last year convinced me I'd never pass. But half the candidates do. Coincidentally, that's the equivalent of the pass/fail rate under the old screen. The difference is the immense savings in reducing the screen from nine weeks to one.

It now costs about $1000 per candidate. Previously, a person was hired as a GS-7, put on full salary plus $42 per diem and run through the course. It was a gamble. A person hoping to break into ATC had to quit his or her previous job and risk being unemployed nine weeks later. Under the new screen, you take a week's vacation and, if you fail, no harm.

No job yet

The half who pass are put on hold. Eventually, a class becomes available and the new hire is put on payroll (GS-7) and sent, at government expense, back to Oklahoma City to begin training. From that day forward, the job title is Air Traffic Control Specialist. The emphasis at the FAA Academy—and in the field—is to "train to succeed." In other words, the FAA is putting out a lot of money to turn this person from an ordinary citizen into an air traffic controller so no effort will be spared to achieve success.

Frankly, this is new and promising. Those who have been around the FAA for many years look back upon their training as a rite of passage, a thoroughly miserable time during which it was made abundantly clear that no one really wanted you there. Training was more of a "let's see what you can do, knucklehead" experience than a time to learn. One became a journeyman or full performance level controller (FPL) after being passed through many phases of training, each worse than the previous.

Today, under the kinder and gentler train-to-succeed philosophy, the new kid isn't even called a student. They are "specialists in training." Sounds like a road sign, CAUTION SPECIALIST IN TRAINING NEXT 50NM. This philosophy hasn't quite reached all corners of the system, however. Being a "specialist in training" is still the draining experience it was when I was a developmental controller.

What does the aspiring controller achieve after all the aggravation? Depending on the facility, the paycheck ranges from GS-10 (base salary $21,906) to GS-14 (base salary $54,607). With overtime, night differential, holiday pay, Sunday pay and a few other premiums, a senior controller at O'Hare can make more than $80,000. Average lifespan, however, may fluctuate.

A GS-12 controller at a mid-level—or level III—approach control spends about two years in on-the-job training (OJT). OJT becomes the daily work for the new controller. Fresh from the Academy, no time is wasted burying him or her under tons of regulations from the 7110.65, the controller's manual, plus local orders and procedures. A particularly nasty task involves "learning the map." Every controller must memorize the entire air traffic control map for the local airspace. The new controller, after about a week, is expected to draw, from memory: *all* the navaids and their frequencies, *all* the airways and radials, *all* the intersections and mileage, *all* the MEAs, MCAs, MOCAs and every bit of information available. For a facility such as Des Moines, this is intimidating. For the new controller in Los Angeles Center, it's Herculean. Still, few controllers fail the map.

Besides the map, the controller must memorize and draw all the approaches, including missed approach, MDAs, frequencies and holding. Whatever is on the chart is fair game. Of course, this is all stored in short term memory and flushed immediately after the exam.

It's much like studying from a Gliem book before taking a pilot exam. The purpose is to pass the test; learning is coincidental.

Getting rated

Once through the paperwork phase of facility training, the new controller—head spinning with facts—begins one of many lab sessions. In a VFR tower, it may be as simple as reading clearances with the training department's specialist. In a large, complex radar facility, training jumps back and forth between the classroom and the radar room. Programs vary from place to place, but follow a nationally mandated structure. The new controller is working toward a "facility rating," much like a type rating on a pilot's certificate.

A controller in the terminal option leaves the FAA Academy with a Control Tower Operator's license (CTO). It's identical to a pilot's license. Where a pilot certificate states, COMMERCIAL PILOT, a CTO reads, CONTROL TOWER OPERATOR. Under "Ratings and Limitations," my certificate shows Des Moines Tower. Never have I been asked to show it. I guess it's hard to ramp check a controller.

The controller graduating from the Academy is issued a CTO with Academy Tower as the initial rating. Academy Tower is fictitious. It's modeled on the old Pensacola Naval Air Station runway layout; the academy students are taught to work traffic in a simulated environment. They must learn the Academy Tower map with all its airways and intersections as though it were real.

The students are broken into teams, much the way they'll work in the field. Instruction is given on a particular subject followed by lab work where students apply their knowledge. The Academy Tower is a large bright room with two levels. On the raised platform is a mock-up tower cab complete with wind instruments, radios, ILS panels, ATIS, phones, FDIO printer, all the appropriate logs, pencils, pads, binoculars and even a light gun. The only thing missing is the coffee pot.

Three steps below the cab is the airport displayed on a large table top like a Lionel train set, only with runways instead of track. Model airplanes provide the traffic. Each model has a hook on the fuselage allowing the "pilots" to pick them up with long sticks. It resembles Battle of Britain-era pictures of the command rooms. It sounds hokey, but anyone having been through it knows how amazingly realistic it can get.

Controllers use headsets just as they would in the field. Instructors sit behind the controllers to teach and evaluate. Instructors also play the role of Center or AFSS calling for clearances. Everything in Academy Tower is a copy of the real world including position relief briefings.

One controller relieves another while airplanes still move. When a student isn't controlling, he or she steps to the floor and becomes a pilot. This doubles the learning experience and would be great practice for private pilot training.

The tabletop tower is low-tech. Basics are learned there and honed in the Academy Tower Simulator—a full tower cab complete with 210 degree surrounding computerized visual display. Should you ever have a layover in OKC, arrange a visit. The Academy loves to show this off, and they should. It's phenomenal. A 28-foot curving screen is on the far side of the tower glass. A bank of moving projectors are hooked to the computer to create traffic.

The computer has a 260-word vocabulary and recognizes the individual controller's voice pattern. Only correct phraseology works, unlike the real world. "Gimme a short approach!" won't do.

The detail in the simulation is uncanny. With binoculars, you can see the propellers turning on the taxiing aircraft. The controller works either local or ground. Unlike the real world, when the controller talks to these planes, they always comply.

If you tell a pilot to taxi into position and hold in front of a landing L-1011, the airplane will do just that. The results are, of course, predictable and graphic. They blow up with a giant, cartoonish orange fireball. A crash truck rushes out to clean up the mess.

The object of the Academy is to give the new controller the skills needed in the field. Ideally, the graduate should arrive at his or her new facility and checkout in minimal time, having learned the basics in Oklahoma City.

In the old days, the graduate arrived at the facility a complete blank, merely thankful for having passed the nine-week screen. Learning began in the field. When I went through the Academy, we were constantly chided, "Don't confuse this with real ATC." Today, the opposite value applies.

Home at last

Having completed the Academy training phase, the controller returns to the field and is put back into the classroom. Months of training lead—hopefully—to certification on clearance delivery, ground and local control. Most OJT is given by other controllers, many of whom are reluctant to do so. The altruism of the Academy melts in the field when a controller is told to be an instructor.

"Here, watch this new kid and if she runs two airplanes together you lose your ticket!" The OJT instructor certification process tends to be a bit of a whitewash. Many who teach are never really taught how to teach. It's assumed that any FPL with six months experience can teach.

True, there are gobs of paperwork to make it *look* as though FPLs have been trained in the fragile art, but it's far from ideal. Instruction seminars are being taught, but are notably lackluster. Last year, I snoozed through a two-day course that made a few interesting points but, for the most part, had little to do with teaching teachers how to teach controllers. No one fails these courses.

The instructor/controller is given a 10 percent premium during those hours giving OJT. For a GS-12, that works out to about $2 per hour to risk losing one's license. It's funny that an agency that treats the achievement of a flight instructor's certificate as the nearest thing to glimpsing the Holy Grail can turn a blind eye to ATC instructors. There's a fair amount of lip service to change this but it's slow in coming.

Training quality varies between facilities but most controllers become devoted instructors. Pride, more than direction, accounts for this. Throughout the OJT period, the instructor keeps the supervisor informed of training progress. As the new controller approaches competence, he's evaluated by someone from the training department or another supervisor who can recommend more training or certification.

In the case of the latter, a "checkride" is arranged and the controller is watched by a supervisor during a period of moderate traffic. If all goes well, a temporary certificate is issued, just like a pilot's temporary, and the controller is anointed as a full performance level controller, or FPL. Usually a great deal of cheap beer is consumed the following weekend.

Return to OKC

For the new tower FPL, the journey has only begun. Assuming the tower is collocated with a tracon, the FPL must eventually return to the Academy to attend radar school. Like Academy Tower, the Radar Training Facility (RTF) simulates an airport approach control environment using the same runway as Academy Tower.

Once again, the maps are learned. Again, the student attends classes and applies the radar skills in a radar lab. Training in the lab is on a one-on-one basis. When I visited last fall, a group was running a problem. The atmosphere was relaxed. Instructors worked shoulder-to-shoulder with the students.

"Ghost pilots" played the roles of traffic from keyboards in another room. Each problem ran about 50 minutes with varying degrees of complexity. Like any simulator, RTF introduces rules and procedures in small pieces. "Chunk and apply" it's called; learn a chunk, then apply it.

Throughout the 21-day program, complexity is increased as the rules compound and traffic volume builds. Although this portion of training is a pass/fail situation, the academy's goal is to pass students by teaching what they need.

It's not a place to turn on the pressure and make or break students. That, hopefully, took place in the screen. "Train to succeed" is honestly applied here.

Clubhouse turn

Leaving the Academy for the third time, the controller returns to his facility to begin another phase of OJT. Again, the student is taken through a classroom to review airspace and procedures. If a simulator is available, a week or so of problems, tailor-made for the facility, will be run. The Academy should have taught the controller the basics of separation, speed control and phraseology, leaving the local facility to teach the particulars.

In the rosy future, the FAA Academy is envisioned as so efficient as to graduate students ready to jump into level III traffic with little more than a quick briefing. That may be overly optimistic. It's much like taking a student pilot through a $24,000 pilot course then plunking him into the left seat of a Cessna 421 and saying, "Lots of ice up there. Watch the thunderstorms and good luck!" Nothing replaces sweaty-palm experience.

The controller-in-training, regardless of how well he or she may have done in the controlled environment at OKC, still must train under the mentorship of someone named Frank who has been vectoring since the Kennedy administration and may not view the new training initiatives with the same enthusiasm as the student or management.

Like flight instructing, the OJT process will always be reduced to the relationship between student controller and FPL. Training will always be a stressful experience. Once completed, however, the new FPL can relax, pick up bad habits and begin to swagger. Six months later, the controller who thought her career was in the toilet may find herself, clipboard in hand, greeting a new kid from the Academy.

Controller review

Every ATC facility has a training department. These vary in size from one QATS (Quality Assurance and Training Specialist) to the larger staff rooms in a big Center. The training department keeps the orange folder (FAA form 3120-1) that follows every controller from career cradle to grave.

The QATS is usually an FPL controller working his or her way up the career ladder before making the leap into management. Most QATS are quite good, both for reasons of personal job pride and the desire to make their bones in a boring, underpaid staff job in hopes of getting noticed.

Besides guiding new controllers through the training program, the local QATS must administer the recurrent training for FPLs. This can be tougher than teaching the newcomer. FPLs tend to resent being critiqued. Each controller must undergo numerous briefings on changes to regulations and orders. Quarterly, controllers review a variety of rules that may not be commonly used and are possibly forgotten. This includes emergency procedures, lost aircraft procedures, hijacks, Presidential aircraft handling, snow removal, thunderstorms and other severe weather techniques.

While the FPL doesn't have to worry about passing these classes, some of the material may prove important. All training is logged in the controller's folder and initialed by the controller. It's much like watching Col. Blake from M*A*S*H initialing everything Radar O'Reilly slides in front of him. The training folder provides evidence that, yes, the controller knew about VOR orientation so should not have run that lost pilot into Mount Rushmore.

Twice yearly, the controller is given an "over the shoulder" review by a supervisor. The supe leans over the controller's shoulder and tries to keep up with traffic. Notes are made of procedural errors. It's similar to a Part 135 ride only no one "pulls a scope." Also included in the recurrent training are tape talks. Controllers must listen to randomly selected pieces of their own work. This can prove embarrassing; "I sound like *that?*"

Where available, problems are posed that require the FPL to transition from radar to non-radar procedures. Since non-radar is rare, controllers aren't very good at this—the minimal procedures are established and designed for survival rather than proficiency.

Recurrent traffic is often a ho-hum exercise followed by a lot of initialing. Occasionally, a controversial item will come down from

Washington and get everyone's attention. Often, the QATS hasn't been supplied a clear explanation of the change and the briefings degenerate into animated free-for-alls. Eventually, everyone throws up their hands and agrees that the Washington crowd hasn't a clue as to how air traffic is worked. It's much the same rule implementation process used for a new FAR. A rule is tossed into the crowd, then the rulemaker stands clear until the dust settles.

For the pilot, an instrument competency checkride or BFR can be a learning experience or something administered on a barstool. The value of each is obvious. The same applies to the controller. A good pilot is the product of good instruction. A good controller probably learned from another good controller who learned to teach despite the system.

Not for everyone

Obviously, this job is not for everyone. Now that you've seen how we get to be controllers, you'll realize that we're not really short-tempered bridge trolls with sadomasochistic tendencies...we just act that way.

Seriously, it is a great credit to FAA's methods that the caliber of controllers the Academy turns out is consistently excellent. If you recall the PATCO strike, there were dire predictions of havoc in the ATC system because of the firing of so many FPL controllers. Didn't happen, and the reason is the quality of controller produced by the FAA.

In the Hot Seat: The Life of a Controller

Few of us have the slightest notion of what it's really like to be an air traffic controller. We know it's stressful (why else would these people be so tense all the time?), and that the job involves sitting in a dark room and staring at a big round radar screen. But beyond that, we know as much about what goes on at ATC as our non-flying acquaintences know about what it's like in the cockpit: you can explain, but there's no substitute for being there.

In this chapter our controller contributors offer some rarely seen insight into what they do for a living. First is tracon controller Paul Berge with a story about that creature we were introduced to in the last chapter, the controller trainee.

The Watchful Eye

Ever been flying along, head in the clouds, needles centered, carefree as the day your last divorce came through, when all of a sudden ATC's previous clearance is countermanded by a new voice; a slightly older, tired, and possibly terrified voice? Ever wonder who that was?

It's the trainer, the journeyman controller who's been leaning over the trainee desperately trying to decide how far to let his budding colleague go before jumping in and reestablishing order.

It was a Thursday. Things happen on Thursdays. It's the day everyone flies, and all the weather comes in for refills. It was on this particular Thursday that I was asked: "Hey, would you mind watching Ed on west Radar?"

"No problem," I said. "Who's Ed?"

Ed was a trainee. Des Moines Tracon had, somehow, been designated a "training facility" by the three-piece suits who make such decisions. Ed was getting close to his checkride, and pretty well had this ATC routine figured out. My task, as instructor, seemed easy: Watch Ed work traffic. Stay awake.

Des Moines was IFR; ceiling about 800 feet, visibility two miles. The ILS 12L was in use, and we were vectoring to the final approach course—easy stuff.

The late afternoon rush was on with arrivals being handed off from both Minneapolis and Chicago Centers. United and American from the east; Republic Airlines from the north; Frontier's 737 from Denver; a handful of corporate pieces here and there demanding to be first; a sprinkling of A-7s shouting "minimum fuel" and one unobtrusive cargo-hauling Boeing 727 belonging to a firm whose name completely slips my mind. We'll call it Hauler 2192.

Ed had been working traffic for about 20 minutes and was doing a bang-up job. Configured in a two-scope operation, Ed handled all the traffic from the west, and someone training on east radar owned the other half of the airspace. The flow was simple. With the momentary rush, all the arrivals were being vectored to a downwind, descended to 2600 feet and slowed to 170 knots.

Once lined up, all we had to do was negotiate with the other controller for a sequence, then turn the planes to a base leg followed by, "...five from the marker, turn right heading 090, maintain 2600 until established on the final approach course, cleared ILS 12L approach, speed 170 knots."

Listening to Ed crank out the clearances was like hearing the recorded voice on the telephone give the time—unvarying, and exact to the point of being dull.

But Ed had done something a little out of the ordinary. What was it?

"Ed," I asked, pointing to a King Air lagging behind on the downwind, "why'd you slow him to 150?"

"Oh, I dunno, thought it'd be a good idea."

"Well, Ed," I said leaning toward the scope and ignoring Hauler 2192 at 2600 on the downwind paralleling the localizer outbound, "by slowing the King Air, you've..." and I went into my standard spiel, the voice of experience and so on.

Ed kept nodding his head, looking suitably contrite and listening to my speech about speed control.

But hey, wait a second. What's that conflict alert horn blaring about?

Ever do something so dumb, it takes several seconds to comprehend the magnitude of your mistake?

"Ed," I asked cautiously, seeing Hauler 2192 at 2600 feet, angling toward another Boeing 727 loaded with passengers tracking the localizer inbound, "what heading is Hauler on?"

Ed's answer, "Ahh, ahhh, ahhh...," told me what I needed to know. Whatever heading he'd assigned the pilot had dutifully obeyed. It was pointing Hauler directly at the air-carrier 727 going the opposite direction. With a closing rate of 300 knots, these two aluminum gymnasiums full of parcels, passengers and leftover reusable omelets, were about to collide over West Des Moines, which would no doubt result in some embarrassing questions.

"Hauler 2192, turn left a whole lot, vector for excuse, excuse, etc."

"Roger, Hauler's turning to ridiculous heading....," the pilot's bored voice came back.

The two jets, both in the clouds, passed—same altitude, targets barely apart, the conflict alert shouting its lungs out until the two were again a legal distance apart.

I looked at Ed. His eyes were wide; I suspect mine were, too. We both glanced toward the supervisor's desk where a mountain of paperwork was keeping him too busy to note what had almost put Des Moines on the front page.

"Ed," I said past a dry tongue. "Whatever heading you gave Hauler didn't work."

"Uh-huh."

The rest of the session slid into the routine. Ed never uttered that heading, and the rush tapered out to a few Cherokees shooting approaches.

So, what went wrong? Obviously two jets came too close, and the mistake wasn't caught. But why had they done so? Two sets of eyes were monitoring the situation. It was all routine, no severe weather, no emergencies. The conflict alert worked. What was the real gaff?

Me!

I'd allowed myself to be lulled by complacency. Ed *sounded* good, therefore, I assumed he must *be* good. Concentrating on some obscure speed restriction, I missed his bigger mistake of pointing two targets at each other.

The CFII, who nitpicks the little stuff might someday fail to catch a serious mistake such as a blown DH or misidentified navaid. Ditto the ATC instructor who concentrates on the minutiae while ignoring the big picture will wind up in the FAA's Home For Burnouts in Oklahoma City, shuffling along the hallways in a bathrobe, eyes wide and mumbling to himself, "I thought it would work...honest!"

One thing instrument pilots have drummed into their heads from day one is the importance of altitude. Go too far below an MDA or DH and you're dead. Stray too far off an assigned altitude and you're busted. (You know what happens when you're busted...they come and take your first-born, then force you to fly back-course localizer approaches in an out-of-rig Cessna 150 with a 250-pound examiner in the right seat. On a gusty day.)

Believe it or not, the controller is just as scared of getting busted as you are. His or her basic job involves making certain that airplanes are kept apart from one another. If they get too close, the controller is nailed.

It involves separation standards. Every airplane flies along in a cylinder of airspace a few miles across and a few hundred feet thick. The controller may not allow the cylinders to intersect—simple as that. (We'll have more on altitude busts in a later chapter.)

Denny Cunningham, a controller at Chicago - O'Hare, relates what happened to him when things didn't go quite right one day.

The Deal

It was a bad day from the start. A warm front had moved in overnight, bringing low ceilings and visibilities and a light rain that would continue all day. By the time I headed for O'Hare for the 3 to 11 p.m. shift, the tollways were a fiasco. The normal 40-minute drive took more than an hour, putting me into the tower with just ten minutes to spare.

I was working the last day of a six-day week. The Monday morning quarterbacks would later say that this was a factor but I don't think so. I like to believe that anyone who's been working airplanes at O'Hare for as long as I have (more than 13 years) isn't about to be rattled by a long work week.

Whatever. I was there, the airplanes were there; it was time to go to work on what I assumed would be a routine evening shift. Little did I know that it'd be anything but routine and before the shift ended, I'd be saddled with an operational error—a "deal" in controller slang.

Airport's a mess

Bad weather at ORD is able to achieve what banks of computers and platoons of flow controllers can't; it eliminates the peaks and valleys caused by hubbed airline operations. Schedules get so screwed up that there aren't any arrival and departure banks, just an unending flow.

Overflow arrivals are parked in holding patterns far, far away so we tower folks aren't even aware they exist. But the departures get so stacked up on the airport that ground has to issue delay vectors to out-of-the-way taxiways. It can be a real mess.

We were landing on 9L and 9R and departing on 4L and 4R. Sounds simple, but a look at the airport diagram shows a couple of gotchas built into the program, particularly for the controller (me) landing airplanes on 9R and departing on 4R.

For one thing, there's no convenient taxi route for the departures.

Every airplane has to cross the landing runway (9R) on its way to 4R and because of the way the taxiways are designed, just-landed airplanes are slowing rapidly on the runway, meaning that the next airplane on final is bearing down on him. This makes the timing of the crossing clearance critical.

Once a departure has taxied into position, it's tricky to time his takeoff. With the weather as low as it was, a missed approach on 9R was always a possibility. A prudent controller is careful to launch the 4R departure at a time when the airplane won't be off the departure end of 9R in the face of a missed approach.

Breaking rhythm

I'd been on position about 15 minutes, long enough to get into the rhythm of accomplishing tasks in just the right order and at just the right time. When things are clicking, a controller has to be particularly alert when that rhythm is broken.

Things started to unravel when a supervisor gave me a radar pointout on a Navy C-12 (King Air) going into Glenview Naval Air Station (NBU), seven miles northeast of O'Hare.

That base leg puts the airplanes right in the face of the departures coming off O'Hare and the Navy guys turn final a mere five miles east of O'Hare. That's the reason for the pointout, so the O'Hare controller can plan the flow to provide at least three miles lateral or 1000 feet vertical separation.

When I got the pointout, the C-12 was 18 miles east; plenty of time to get another departure out, or so I thought. I cleared a USAir DC-9 into position on 4R, then checked to make sure a 9R arrival wasn't going to surprise me with a missed or a 9L arrival wouldn't suddenly be a factor for my departure. By the time the DC-9 was in position, a glance at the radar showed the C-12 to be about 11 miles east. No sweat.

Plan A, B, C

I cleared the DC-9 for takeoff, with an initial heading of 090, to clear him of a possible missed off either runway 9. My plan was to see how long the -9 took to get airborne, check out his rate of turn and climb and decide at that point how to provide the required separation from the C-12. Controllers, like pilots, always have several plans in mind; each is fluid and flexible, to suit what develops. I had three strategies.

Plan A: I'd leave the -9 on 090, since that was his on course heading. But that would require him to climb well enough to get through 3000 (1000 feet above the C-12) before the C-12 came within the magic three miles.

Plan B: If the climb was looking doggy and the C-12 was getting close, I'd stop the -9's turn on a 050 heading, which would take him well north of the C-12. Once he'd climbed through 3000, I'd turn him eastbound on course.

Plan C: If the climb wasn't a sure thing, but his rate of turn would take him southeasterly without busting three miles, I'd do that. This would take him south of the C-12 until he'd climbed through 3000. Then I could turn him on course.

I was confident that at least two, and probably all three plans would work. I'd done this sort of thing a million times. All I had to do was find out what the -9 was gonna do in the turn and climb.

As the DC-9 rolled, my attention was diverted to my next departure, a 737 that needed to cross 9R. His way was blocked by another 737 and by the time we sorted that out and I'd checked on the DC-9, he was already off and a mile or so east of the field, but still low. The confusion over the 737s had broken my rhythm; I was 30 seconds late in making the decision as to which of my three options was best.

Head on and closing

The C-12 was now head-on and eight miles away. The -9's climb was doggy so I chucked Plan A. Plan B wasn't looking so hot, either. My delay meant that the -9 was already through 050 and would have to turn back left if I hoped to get him north of the C-12. Too late for that, so Plan C it is. I'll drive the -9 southeastbound, get him out of 3000, then crank him back eastbound on course. I keyed the mike.

"USAir 464, continue your right turn, heading 130, give me a good rate through 3000."

"USAir 464, roger."

As I watched the targets approach each other, I felt an unfamiliar and unpleasant feeling in the pit of my stomach. This wasn't working. The DC-9 was climbing like a parakeet with a broken wing and his turn rate suggested that he was trying hard not to spill his coffee.

As the -9 climbed through 2000, it was obvious that a collision was impossible; the C-12 was still four miles away. But even though I'd met the prime directive of keeping the metal from merging, it looked like a separation bust was in the works.

"USAir 464, turn right heading 180, tighten up the turn, hold a good rate of climb."

"USAir 464."

It was useless to issue traffic, both airplanes were solid IMC and the -9 was already above the C-12 anyway. Obviously, they had no idea of the dilemma I'd created for myself or maybe they'd have turned and climbed better than the 1-1/2 degrees per second and 1000 feet per minute that later analysis showed they'd given me. It might've helped if I'd asked them to report leaving 3000 feet. Such a report takes precedence over the Mode-C readout and might have shown the DC-9 leaving the altitude a few precious seconds sooner.

As I watched the targets on the radar, the feeling in my stomach worsened. I called across the tower to the supe.

"Hey, Jack, I think I'm having a deal."

Before he could answer, the voice of the tracon supervisor came over a nearby speakerphone.

"Hey, tower, what are you doing with USAir 464?"

Exactly how close had the two airplanes had been before 1000 feet of vertical separation was established? If the answer was less than three miles, I wasn't going to be talking to any more airplanes for a while.

I keyed the mike one last time:

"UsAir 464, turn left heading 090, and contact departure."

This guy would get no "good day" from me.

Off duty

Seconds later, I was relieved from the local control position, as was the tracon controller involved. After a brief statement, the radar controller was back at work later that day. Not me. I was about to be decertified.

A call went out to preserve the audio tapes; another to the computer guru who would later reconstruct the flight paths of the two airplanes from computer and radar information. I was also offered the opportunity to contact a union representative.

I was given a blank piece of paper, a pen, and a quiet room to jot down any notes for my own use. Shortly, the tape of the tower frequency was played for my listening pleasure. I was startled to hear how long the exchange between the two 737s took; I hadn't realized how much it had distracted me. With a sigh of disgust, I then sat down with the AM, or area manager, a friend of 13 years and who trained me at O'Hare.

"So, what happened? Did you forget about the Navy guy, or what?" he asked.

There were 12 controllers in the cab at the time and since I was his most experienced controller, he'd assumed I'd be able to do my job without his assistance. I was embarrassed that I'd let him down.

Still, at this point, we didn't yet know if it was an operational error. That would depend on exactly how close the airplanes had come. This would be determined by the National Track Analysis Program, or NTAP, a computer-generated picture of the airplane's paths.

We hovered over the computer tech's desk as the printer did its work.

It didn't look good; the paths of the two airplanes intertwined like two loose pieces of spaghetti.

When the machine was done, the AM laid the piece of paper on a desk, got out a ruler, and started measuring the distance between the two airplanes at matching times indicated on the plot.

"Definitely less than 1000 feet or three miles. Sorry, Denny."

"Me, too."

Decertified

I was no longer qualified to work air traffic or, as we say in the FAA, decertified. I'd have to retrain for the position. While I was done for the day, the AM's work had just begun. He notified the appropriate authorities farther along the FAA chain of command.

When I returned to work Monday, the bad news was official. The QA guys had measured three pairs of radar hits that depicted the two airplanes as having less that required separation. Total time the airplanes had spent in illegal proximity to each other: ten seconds.

As deals go, this one had been relatively benign. Yes, it had been a lousy operation but there was no danger of collision. It was illegal but so is doing 65 on the tollway.

The QA guys have their own agenda. They're looking for trends and problems with procedures, training, supervision or equipment. Mostly, they're trying to find a way to avoid having to investigate another such incident in the future.

Since the FAA is not an agency to just say "Stuff Happens," a new rule or procedure usually results. In this case, a new order was issued prohibiting departures off this runway combination whenever an NBU arrival came within 15 miles.

Back in the saddle

My own agenda was to get back to work. I knew what the problem had been. I goofed, big-time. It's been said that air traffic controllers make thousands of decisions every day, and that every one of them has to be right. I disagree.

Yes, controllers do make thousands of decisions but even if they aren't all right, it doesn't mean airplanes collide or separation is violated.

No controller works with minimum separation as a goal, there are too many uncontrollable factors to consistently beat the odds. So all of us build in little margins. In my effort to achieve "safe, orderly, and expeditious," flow, I'd obviously cut that margin too thin. No one knew that better than I did.

But remorse and repentance are not enough to get a decertified controller back on the boards. Policy required a review and even a written test on procedures. My supervisor took longer writing the test than I did taking it, but it was finally done.

After the written test, we trekked up to the tower so I could "demonstrate the practical application...." The entire process was complete and I was talking to airplanes again by Monday afternoon.

Three months later, I received a copy the 16-page report on the incident. In typical FAA-ese, it said "Employee A intended to apply radar separation between aircraft #1 and aircraft #2...Employee A became distracted with the task of identifying the first aircraft holding to cross runway 9R...When Employee A again turned his attention to aircraft #1...separation was lost."

Speaking as Employee A, I can't disagree with that conclusion, although I'd have probably rephrased it:

"Employee A momentarily had his head up and locked, causing him to put aircraft #1 and aircraft #2 closer together than appropriate. He told aircraft #1 to zig, when he should have told him to zag."

I guess it just goes to show you, if you ain't payin' attention, stuff happens.

The last arrival blocked in an hour ago. The frequency and interphone have been silent for longer than that. The only sign of movement on the airport is the yellow beacon on a city truck prowling around the fuel farm. Otherwise, the view out the cab window is nothing but blue taxi lights floating in a sea of black. A better sleep-inducing view is hard to imagine.

Welcome to the mid, ATC's version of the graveyard shift. Some controllers actually like the mid, which variously runs from 11 p.m. or midnight to just after dawn. They say you get used to working when everyone else is sleeping and when every bone, muscle and synapse in your body says you should be doing the same.

Like it or not, though, the mid is a grind. Even though that voice in your headset may sound alert and awake, strange things happen in the lonely hours before dawn. These next four stories—all true—were related to us by controllers we've become acquainted with during our travels. The names and runway numbers have been changed to protect the innocent.

Who's That on Final?

Traffic at a busy terminal often goes well into the night, but on this midnight shift, the last rush of the night was finally gone. The problem now was to stay awake, which would be particularly tough since it was a Saturday night and traffic was at its lowest ebb of the week. It was likely that no airplanes would take off or land at the airport for the next several hours.

There were no proposed departures in the strip bay and besides, they'd be calling clearance first. The first notice the controllers would have of an arrival would be the pilot's initial radio call, 5 to 7 miles from the runway. Although the ARTS tag of any arrivals would appear on the tower BRITE at 20 miles, who looked for ARTS tags at two o'clock in the morning?

With traffic nonexistent, the airport had closed all the runways for maintenance, except the active, 23L. The yellow flashing lights of half a dozen maintenance vehicles were scattered around the airport as holes were patched or light bulbs replaced. The quiet in the tower was broken only by the low voices of its only two occupants; a supe and a controller, who were debating what kind of computers to buy for their kids.

Neither is sure what first drew their attention outside the cab. The supe thought it was a radio call from an airline taxi crew that caused them to glance out the window. The controller thought he'd first noticed the ARTS tag on the BRITE scope. Whatever, the results were the same, both gaped dumbfounded at the sight of an Air Cargo Boeing 747 on short final for Runway 23R. The lights of two maintenance trucks flashed dimly on the closed runway.

The supe got to the mike first, scanned the BRITE for the callsign, then keyed up: "Air Cargo 78 heavy, go around!" No response. As he keyed the mike to try again, the controller picked up the interphone line to the radar controller, sitting alone 200 feet below and yelled, "Tell Air Cargo 78 to go around!"

The tracon controller, obviously noticing for the first time that the blip on his scope was lined up for the wrong runway, said defensively, "Hey, I cleared him for 23L."

All pretext of the cool, calm, always-in-command controller evaporated as the tower controller yelled, "Tell him to go around!"

The radar man finally keyed his mike and gave the instruction. The 747 pilot didn't answer immediately, but he got the message. The heavy was well inside the threshold and within seconds of touchdown before the descent was reversed. The city worker in charge of one of the maintenance crews working on the runway could hardly be heard over the roar of the jet passing a few feet overhead as he frantically inquired as to whether the tower realized the runway was closed.

The after-incident phone call from the captain was a litany of contributing factors. The crew was used to landing on 23R so that's what they did, out of habit. The copilot flying was new and well behind the airplane; although they'd been cleared for a visual, they'd become preoccupied with the fact that they weren't receiving the localizer. And yes, they'd received the instruction to contact the tower, but had become distracted with the localizer problem before doing so and had simply forgotten to switch to tower.

The supervisor, besides having no fondness for paperwork, also realized that the ATC personnel involved would take some of the blame for not catching and correcting the mistake. He assured the pilot that since an air carrier go-around was not an unusual event, there would be no repercussions and that the incident was closed.

He thanked the pilot for calling, hung up the phone, then gazed silently out the window at the trucks still working on 23R. With a sigh, he turned to the controller, and his voiced trailed off as he said, "Okay, I think my pulse is back to normal; why don't you go get us a couple of those NASA forms...."

Hey, This Doesn't Look Right

So there I was, impersonating a (yech!) approach controller at 3 a.m. in a Center radar room on the west coast. It was my first mid shift in a few months, so I was a little out of practice.

A nearby approach facility had packed up hours before, abandoning their airspace to us until 6 a.m., leaving us with four of their airports to take care of. A Seneca checked in VFR, requesting the ILS 28 to one of those fields.

I started the usual patter, issued the altimeter setting, and rummaged around for the Gummint issue, occasionally-not-even-expired NOS approach book kept above every Center radar scope behind the coffee cups and gum wrappers.

Now, imagine going into a French restaurant, ordering something relatively complex and then spotting the chef through the kitchen door stirring with one hand and reading a cookbook with the other. Well, I was doing the ATC version of that, reading the approach plate to figure out how to handle this guy while simultaneously talking to both him and the FSS on the field and trying to sound smooth. Lucky for controllers, radar isn't a two-way mirror. We were stealthy long before it was fashionable.

The normal transition to the ILS 28 approach was via a 22-mile DME arc to join the localizer. But I decided to be a nice guy and do some fancy vectoring instead of having him fly the full approach. The plan was to put the aircraft on the final approach course at about 5000 feet and 20 miles from the airport.

Now, what to give for the required crossing restriction? Well, between the arc and an NDB located on the localizer about 12 miles from the runway, there was a note which said "5000 to the NDB." That meshed well with my 5000-foot minimum instrument altitude so I herded the Seneca around to the final, ending with, "Fly heading 250, intercept the ILS 28 localizer, cross the beacon at 5000, cleared ILS approach." The pilot rogered the clearance, reported established on the localizer a couple of minutes later, and was sent off to talk to the FSS man, who was reporting the weather as 300 and maybe 3/4 mile in fog.

I sat back to watch the Seneca's progress and was feeling quite satisfied with myself for doing such a good job. Then I saw him whistle over the airport at about 1100 feet. Huuuh? Whatsamatter here—can't follow a glideslope?

Seconds later he was back, reporting a missed approach and asking for another go at it. Being ever so obliging, I brought him back around, issued the same clearance and sent him back to the FSS. This time he got in. Then I noticed the problem.

Despite the "5000 to the beacon" note, the GS altitude was shown as 3687 MSL, a good 1300 feet lower. I'd put the poor guy about a quarter of a mile above the glideslope, twice, with near-minimums weather, and thoroughly on purpose. I know this is the sort of thing pilots think controllers do for amusement ("Hey, Chuck, I got this Seneca to fall for that Kamikaze arrival! Come watch!"), but in my case it would have to be called an honest mistake.

In retrospect, I was amazed that the pilot didn't object as he watched the glideslope needle rocket from full up to full down as he bored toward the beacon at 5000 feet. It must have been an interesting approach.

A couple of days later, I called the Sacramento Flight Inspection Field Office and had a little chat, during which I allowed as to how there seemed to be something wrong with this approach. Shortly afterward the note was changed to "5000 to GS intercept," which made a *lot* more sense. I never did anything dumb like that again. (Subsequent screwups were completely different!)

Listen Up!

The controller had been awake for more than 24 hours and was three hours from finishing the shift. Four a.m. Snowplows crawled around the terminal ramp, trying to push away enough snow for the airliners to get out. Ozark had been stranded since noon, and United was being handed off from Center, six hours late.

It had stopped snowing around midnight. One runway open; braking action poor. Two connecting taxiways and a parallel taxiway were all that led to the ramp; braking action poor to nil. Ozark pushed back and crept toward the runway. It looked like a fat lady trying to ice skate.

A BE-18 full of Wall Street Journals called ready to go and was cleared for takeoff. It disappeared into the clouds about the time the controller called radar contact. The same controller worked clearance, ground, tower and approach; a standard, after midnight shift. One controller, all positions, no computer interface, no help, no sleep. No joke.

It wasn't that traffic was so complex. The problem was working with a fuzzy brain. The FAA spends millions testing controllers for drugs but never for sleep, or lack thereof.

Ozark called ready. The controller was talking to the snowplow operator on the telephone, taking a long-winded field condition report. "Stand-by," the controller answered, telephone pressed between his shoulder and cheek, microphone draped across his wrist.

United asked if he should intercept the localizer as the Twin Beech asked to turn on course. "Yeah, United 123, cleared for the approach..." The BE-18 repeated his request and was cleared on course.

A Cessna 210 called for taxi from an FBO not yet cleared of snow. The drifts didn't seem to deter him.

Ozark sat. A snowplow asked to make a turn around on taxiway B at runway 29R. United was cleared to land, and the controller told the snowplow coordinator he'd have to call back for the rest of the report. He didn't really give a damn about it anyhow.

The plow reported clear of the runway and the BE-18 gave a tops report, which the controller rogered, wrote on a strip and promptly forgot.

"Is United cleared to land?"

"Cleared to land...braking action poor, taxiways poor to nil."

The Cessna 210 called ready to go. A YS-11 called ready to taxi from the cargo ramp; said the ATIS wasn't on the air and "what's the wind?" A Northwest DC-9 called for taxi and verified that the ATIS was off. The controller needed to cut a new one but first, where was Northwest's clearance and what'd that 210 want?

The controller read the wind, altimeter, paused and read the latest weather, which he noticed was an hour old. It was harder to focus than it had been an hour ago. He wanted to sleep, just a while.

Ozark asked politely when they could go.

"Ah, hold short...traffic on final." The controller picked up a phone and called National Weather Service for the latest weather and was told it was sent, "...must be something wrong with the printer."

United popped out of the overcast as a snowplow driver asked to cross the runway. The controller started to tell the plow to cross and hesitated, then, seeing United, said, "Hold short." There was a pause and someone keyed the mike and said, "You know, tower, we could've made it out there!"

Until then, the controller had been working in a fog of caffeine-fired nerve endings. He wanted desperately to lie on the floor and sleep. He wanted the airplanes to go away. He wanted the airport to close until the snow could be removed, until the day shift could open the tracon and someone else could take ground and clearance and maybe a supe could take the damn inane calls from the city about snow removal.

"Ozark," he said in a tight voice, "I'll get you out when I can." Ozark was quick to answer: "That wasn't us who called. We're in no hurry."

Another pause, then 24 hours of no sleep took the controller's tongue: "May I have everybody's attention, please! I've been awake for going on 25 hours now. I've worked two shifts. I'm tired. My eyes are crossed and out of focus. The computer's down and I'm working all positions here. I've got six aircraft, four snowplows, a couple of pick-up trucks, one runway; the braking action stinks and I have to go to the bathroom. We're doing this my way or not at all!"

Silence.

United jumped in: "Wasn't United, tower!"

One by one roll call was held: "Northwest didn't say it!"

"Twin Beech 51X didn't either!" Each pilot keyed the mike disavowing culpability; each pilot except the Cessna 210. Ever seen a coyote cull a sick lamb from a herd? I haven't either. But I suspect the Centurion pilot must have felt awfully lonely sitting on that snowy taxiway watching every aircraft taxi past him and depart, until he was the only one left, a pathetic dot beneath the FAA tower.

"Centurion 321, now are you ready?"

A cooperative pilot, realizing he was being controlled by an ATC gone over the edge, decided to taxi back to parking. He probably took the bus.

Anybody Home?

According to the flight progress strip, the Cessna 310 should have landed. It was a routine flight—a freight hauler who came in every night, same time, same route, same speed, requesting the same downwind landing to make his connections.

Only this night he just flew past, as though he'd read too much Richard Bach and decided, "I don't want to be a freight pilot, anymore. I just want to fly...and fly..."

And he did. He flew right over the airport. I tried to call; no answer. I even tried other pilots, who yelled at him to answer approach, to no avail.

It was a clear night; 2 a.m. Working all positions alone in the tower cab, I could see the 310's target move slowly across the scope, as well as see the blinking strobes pass 5000 feet over the airport. Good-bye, sleeping beauty. Good luck!

I called Center. "Hand-off. Don't know where he's going. Don't even know if the pilot's alive."

Center shrugged. "We'll watch him."

About 50 miles north of his stated destination, the light bulb came on. The Twin Cessna's target wavered, then circled, then pointed itself back at the airport. "Approach, Laidback 612, with you, ah, got the ATIS, ah, airport in sight... (sotto voce: 'where the hell am I?')."

I've often fallen asleep on the midnight shift. It's a little startling to be seated in front of a radar scope, watching a target on the east side of the scope and suddenly discover it's moved to the west side without any idea how it got there. My brain wants to sleep at night, not stare at radar. I can only imagine how that 310 pilot felt popping awake and having no clue as to his whereabouts.

So, if you're flying along some night, needles centered, DME ticking off the mileage and no one seems to be home at approach, remember this: It's not that we've closed up shop without telling you; we're just sleeping. Land gently, taxi quietly and watch for traffic. Good night.

Not all ATC disasters and tieups are caused by weather, too much traffic, equipment problems, or even fatigue. Sometimes all it takes is access to the computer and just enough knowledge to be dangerous....

The Perils of Databoy Dan

At the lowest link of the evolutionary ATC food chain, exists a life form virtually ignored by the flying public. Barely paid a subsistence wage, these bottom feeders are lumped together under the acronym: ATA.

Air Traffic Assistant; it even sounds bogus—not a real air traffic anything, but an assistant, like something you'd learn in a school from a matchbook ad.

"Goodbye, Mom! I'm off to Air Traffic Assistant School!"

"Oh, if only he'd become a *real* controller like his cousin Eb!"

ATAs are the ATC equivalent of gofers. Nobody knows just what they're supposed to do and, as a result, all the undesirable work is dumped on them.

"Who wants to give a tour?"

"Not me! Get one of the ATAs."

Then one night, the ATAs rebelled. Actually, *one* ATA rebelled. And it wasn't so much a rebellion as a simple gaff that managed to shake the air traffic system from Denver to Washington.

His mother had named him Donald. We called him Databoy Don, a simple GS-7 with demonic powers. Databoy was the best ATA at Des Moines and was being rewarded by being promoted into the ranks of real controllerdom.

It was early one Monday morning—4 a.m. to be exact—and Databoy was working his last graveyard shift prior to being sent to boot camp in Oklahoma City. The last rush of late-night check haulers had departed and the first air carrier wasn't due in until 5:30 a.m. Databoy was bored and thought he might compose a farewell message for his replacement, scheduled to work the morning shift.

Being careful not to wake the real controller, Databoy began typing an ersatz flight plan into the computer. He gave it a callsign: N628D. He made it a Bonanza, BE35/R; 150 knots and showed a proposed route of flight from DSM to CID, 80 miles away. He filed a p-time of 8 a.m. local, the heart of the Monday morning rush. As an afterthought, he entered into remarks: "BRAIN DONOR ON BOARD" and giggled when the computer accepted it.

Well, he thought, if one fake flight plan is funny, two would be funnier yet, and three would be quite the chuckle to his replacement, who would be left to wonder why no one ever called for these clearances.

Strange things happen inside an ATC facility at 4 a.m. (On a midshift, I once saw Elvis taxi an SR-71 past the tower.) Sixty-four flight plans later, Databoy pushed back from the keyboard, certain he'd give his relief a real belly laugh at 8 a.m.

At 6:30 a.m., Databoy rubbed the sleep from his eyes, briefed his relief, "Oh, nothing unusual...; checklist complete," and left the tower. By 6:57 a.m., he was home in bed, asleep. At 7:30 a.m., Atlas shrugged.

"This thing won't stop printing!" the new ATA stammered to his supervisor while tearing reams of flight plan strips from the printer. "Look at them all!"

A disconcerted supervisor waded through the flight plans. Pilots called for clearances and were told to "standby; clearance on request." Flight plan after flight plan spewed from the printer. Lear after Lear was filed to Denver. A half dozen Citations wanted to go to O'Hare; another eight to Midway and three to Meigs. Eleven F-18s filed various round-robin flights from DSM to Canada to east Texas.

Clearance delivery ran out of room to post the strips and began lining them up on the floor. Pilots were told to continue standing by. Ground scanned the parking ramp with binoculars searching for callsigns to match the proposals. They came up dry.

The printer ribbon snapped.

Center called wanting to know about all the proposals. They said they'd held controllers on overtime to handle the anticipated departure rush.

The printer kept flailing around until its ribbon became tangled with the paper, making a gagging noise, like a zucchini in a garbage disposal.

Flow Control called and slapped a ground stop on everything off Des Moines. They called O'Hare, and after explaining to them where Des Moines was, implemented flow restrictions westbound from O'Hare and did the same for Denver departures eastbound.

Databoy Don slept the sleep of the unsuspecting.

By 8:32 a.m., with the system locked in uncertainty, flights delayed and tempers short, someone noticed the remarks on N628D: "BRAIN DONOR ON BOARD."

Something wasn't right; something didn't add up. Slowly, the realization was dawning that some sort of virus had infiltrated the system and as pieces were fit into the puzzle, the finger pointed to a slumbering GS-7 who had just trashed the most sophisticated transportation system in the world.

The Russians once had a 17-year-old kid in a Cessna 172 land in Red Square causing profound political and administrative ramifications. Had Databoy Don been at the controls, they'd have surrendered.

Anatomy of the System

You're a rare and lucky pilot if you've never suffered the aggravation and inconvenience of having ATC lose your flight plan. Most of us blame some inscrutable force when this happens (The Ozone—as in "Lost in," or The Computer), but usually the reason is much simpler and more mundane. In reality, the process that starts with you filing a flight plan and ends with a clearance being read to you is remarkably robust and flexible...and it (almost) never screws up. It also doesn't work the way your instructor said it did: for example, the two-hour "window" that you supposedly have before your flight plan vanishes isn't the way things happen in the real world.

Controller Paul Berge now takes us on a tour of the innards of the flight service station, and sheds some light on just how and why your flight plan works...or doesn't.

All About Flight Plans

Working clearance delivery in an ATC facility is like being the receptionist for a Hyundai dealership in New Jersey. You're the first person the client encounters and the least able to solve any problems. Clearance delivery, often staffed by noncontrollers (Air Traffic Assistants or ATAs), reads clearances and that's about it. Real air traffic control begins sometime after you crack the throttle and head for the runway.

Clearance delivery gets the paperwork from the computer to the controller, making certain (you hope) that the pilot has been pointed in the correct initial direction. If a flight plan isn't on the FDIO printer at the delivery position, there aren't too many other places to look.

Explaining to a Falcon pilot that her flight plan just isn't there and that she must now explain to her boss why their $10 million dollar jet can't take off is as embarrassing for a controller is it is for the pilot. Although flight plans don't get lost every day, I've seen enough of them disappear to make me wonder why. Last winter, I decided to visit our local Automated Flight Service Station to see exactly how they process flight plans.

We have the technology

It's not like ten years ago when Flight Service was the wooden shack beside the gas pumps staffed by a guy named Ed who had been there since 1947 and who knew every pilot by first name. These days, AFSS facilities are located in economically depressed communities where a fat ladle from the Congressional pork-barrel brought together millions of dollars of federal funds and dozens of disgruntled guys named Ed from decommissioned smaller facilities. Fort Dodge AFSS is the facility we use at Des Moines. It's located about 90 miles northwest, actually on the field at Fort Dodge airport.

I don't know about other AFSSs, but FOD was designed by German architects now employed in Baghdad in bunker construction. A low, stone-cold building devoid of glass, it sits out of sight of the runway as though it had nothing to do with aviation. Outwardly, it lacks the old friendliness I'd remembered. Inside, on the other hand, the place was impressive, but when you fly an Aeronca, *everything* impresses you.

Each specialist sits in a cheaply manufactured chair in front of a pair of CRT screens. One, the Kavouras, displays chart weather, such as the weather depiction chart and winds aloft data. With a series of one-handed keystrokes, the specialist can flip through everything once hidden inside reams of paper. Beneath the Kavouras was another scope, part of the "Model One," the computer system that's the the heart of the briefer's world. I asked for an IFR briefing from FOD direct MCW.

The briefer walked me through the process. As I gave him my callsign, ETD and requested route, he typed it into a temporary "mask" on the screen. Suspending that for future reference, he began the briefing. Everything I wanted was at his fingertips—SAs, FAs, FTs, PIREPs, NOTAMs. If I interrupted with a question, he'd hit a button, and my requested information was ready, instantly.

Briefing complete, he asked if I was ready to file the flight plan, a question he's not *required* to ask. I said yes, and he flipped back to the suspended information already entered on the flight plan mask.

The mask is an easy-to-read screen set up logically to accept a flight plan read verbatim from a flight plan form (FAA form 7233-1). The pilot gives the information and the briefer types in each item—in order, no jumping around.

If an item is unavailable, such as pilot's address, the specialist can skip to the next entry. Once complete, it's a matter of entering the information, and the flight plan as filed. So where does it go? Straight into a giant flight plan filing cabinet.

P-time approaches

The proposed flight plan, once accepted, sits in an inhouse proposal list. Each flight plan has a proposed departure time or p-time, (not to be confused with random drug-testing schedules). The flight plan remains in the AFSS computer until a preset number of minutes prior to the p-time, say100 minutes.

Each Center host computer has a different time parameter for taking flight plans off the AFSS computer and moving them into Center storage. This number is not terribly important to the pilot, as your flight plan can be found regardless of where it's stored, and any flight plan filed within that time frame will instantly be sent to the Center.

So, 100 minutes prior to the pilot's proposed departure, the flight plan is automatically flushed from the AFSS computer and sent to the host computer. In FOD's case, it goes to Chicago Center's computer, 400 miles away. This is where most users assume it's sent by truck, but actually, it travels over a dedicated line, straight to the Center, where it again sits until 30 minutes prior to the p-time, at which point it's sent to the appropriate ATC facility.

The appropriate facility may be an approach control, a FDIO-equipped tower or a Center that handles clearances for the airport you happen to be departing from.

For my sample flight plan, the data went to Minneapolis Center. Thirty minutes prior to my departure, the Center controller at Sector 28 has my flight plan and throws it straight into the garbage can. Actually, they don't do that; they save it until I call off the ground for it, then they tell me my radio's unreadable and to go home and get a real airplane.

What could possibly go wrong?

So, what can go wrong? Very little, I was initially assured during my tour, and, judging from the efficiency of the briefer, I believe him. Hundreds of flight plans are processed daily at each AFSS, and only a handful fall through the cracks. Of course, when you've passed through that crack three times during the same week, you may think less of the system.

The first obvious problem I noticed was the possibility of the flight plan information never being entered—human error. If you're filing by phone, there's really no way for you to know, without asking, if the computer accepted your flight plan. At 4 a.m., a lone specialist might give you a briefing, take your flight plan and, before entering it, become distracted by another call.

The easiest distraction could come from the pilot. Let's say you've given the briefer your flight plan information and just as it is about to be entered, you say: "Wait, I might need an alternate! What's the weather at XYZ?" You get it; XYZ is W0X0F.

"How about ABC?" W0X0F. On and on, until you say, "Ah, forget the alternate; I'll stick with what I gave you. Bye."

You hang up, thinking your flight plan is in, when a tired briefer might inadvertently clear the screen, simply forgetting to enter it. The flight plan vanishes. Later, when you call clearance, they won't have it; no one will; it never was. You'll have to refile. It's a simple mistake in a computer environment, and although it may happen less than .1 percent of the time, you could be the unlucky one. How do you avoid this? You could ask the briefer if the flight plan was accepted. This will automatically flag you as paranoid, but at least you'll know. You could also request the briefer's operating initials if you're out for blood later, but frankly, this won't get you anything except the possible satisfaction of hanging a briefer; you still won't have your flight plan.

Like real ATC, everything you say on the phone or radio to flight service is recorded. Tapes are saved for 15 days. If any problem arises, the briefing can be replayed, and the reason for the problem quickly isolated. Each keystroke the briefer makes into the computer is likewise recorded in the center computer and is accessible by the FSDPS, Flight Service Data Processing System. If you really want to track down a problem, the information is available.

Once your flight plan leaves the AFSS computer, the information is stored for 24 to 48 hours in a "DD file" or "Dead File," a place where flight plans go to die. If any problem arises with your flight plan, flight service can access this information through FSDPS and hopefully isolate the problem to either Center's computer or the ATC facility where the flight plan should have gone.

The DD file also displays the briefer's operating initials, so they can deal with him directly. Usually, though, it's just to easier to refile. This file, by the way, applies only to IFRs. VFR flight plans are handled somewhat differently.

How about Fast File?

A FedEx pilot once told me, his face reddening, what he thought of Fast File. "I never use it; something *always* gets screwed up." This seemed to reflect the opinion of several other pilots I've asked.

Fast File should be the neatest thing in aviation since courtesy cars. As with courtesy cars, however, you must be patient; read the flight plan carefully, speak slowly and use as much English as possible.

FOD AFSS once had a Fast File message: "Huh, huh, huh (heavy breathing; the pilot had been running). Close out my flight plan; I can't get the damn airplane started! Click." No name was given; no callsign; no airport, phone number, nothing; just some out of shape pilot wheezing into a recorder.

Fast Filers often omit a phone number where they can be reached for 15 minutes after they file. If there is anything unclear on your message, the AFSS will need to call you.

A pilot who mumbles gives the specialist the chore of replaying the tape repeatedly until all the information can be gleaned. It's like listening to a garbled ATIS tape; the listener has to strain for the information and may get it wrong.

Any time a Fast File call comes into the facility, a steady red light illuminates at each position. A specialist assigned to the flight data position watches for the light and takes the flight plan off the tape. During my visit to FOD, the longest any Fast File waited to be processed was about 60 seconds. If you regularly use Fast File or have problems with the system, I suggest you pay a visit to the local AFSS and have a talk with the supervisor. Let him or her show you the system and explain your beefs. Believe me, the FSS folks will try to straighten things out.

As usual, it's the pilot's fault

While the old teletypes may be gone the way of the wooden FSS shack, all flight service stations are still connected via "Service B," a communications line allowing one facility to communicate with another anywhere in the country. When you file your flight plan, the computer automatically sends your flight plan to the appropriate Center computer for later processing. All stations are clumped together in small groups of "families" and assigned to Center computers. A flight plan filed with a AFSS in Sacramento for a proposed flight from PHL to BOS will (theoretically, at least) automatically be sent to the appropriate Center computer. In years past, a specialist entering the flight plan had to plot where he thought your flight plan should go and address the information to the appropriate Center. In those days, flight plans went astray far more often than they do now.

Drop-outs

Of all the ways a flight plan can vanish, the most annoying and universally misunderstood is the time out or, as it's known in the Centers, the drop time. Drop times are the preset periods Center computers hold a flight plan after the p-time has passed.

At Des Moines, for example, flight plans are available for two hours beyond p-time. At two hours and one second, a brief message from the Center computer is flashed to clearance delivery: REMOVE STRIPS N628D. Usually, this happens just as N628D is stowing the gear and tower is shipping him to departure. This results in a lot of scrambling to input a new flight plan prior to handing off the plane to Center. If you've ever wondered why you received a new squawk immediately after takeoff, this may be the reason.

You may have been led to believe that two-hour drop times are standard but it ain't necessarily so. In fact, drop times vary widely throughout the system, and it's nearly impossible for a pilots to know how long the flight plan will survive beyond p-time.

Despite the FAA's insistence that the agency is one big family of dedicated, interconnected workaholics, the truth is what's good for Des Moines would bring laughter in Miami. In fact, just about *anything* related to Des Moines brings snickers in Miami.

Each Center sets its own drop times, and, yes, there is no set standard for doing this. Each Center has a woman named Flo who randomly changes drop times when things get a little too clogged. So, if Miami Center was using 90-minute drop times last time you were there, you may find yourself planless after 60 minutes on your next visit. Sometimes, they go the other way. When a Center is extremely busy, three or four hour drop times aren't at all unheard of.

How are you expected to know what the drop time is? Without calling the Center, you can't. Solution: If you want to be sure of avoiding hassles, call AFSS to update p-times if you're planning to be more than 45 minutes late. If they don't have the flight plan anymore—already gone to Center—they can make the appropriate calls.

You can do the same thing with clearance delivery. Call them on a handheld from the airport coffee shop if you want. "Hey, clearance, Six Two Eight Delta, please keep my flight plan open; my passengers haven't left the bar yet." Clearance may forget, but at least you'll have someone to blame.

Correct time, please

A common mistake that is certain to send a flight plan into purgatory is getting the p-time wrong. Everything is in z-time, Zulu time, coordinated universal time. It is too easy to give local time when filing or to miscalculate Zulu and find your flight plan has been shuttled into the proposal list for 23 hours after you've requested. This happens a lot during the spring and fall time changes, when everyone gets confused about how Zulu relates to daylight savings.

If this happens, don't panic. Leave panicking to ATC; we're paid for that.

If you suspect you've goofed, call AFSS and say "Please check my flight plan; I meant to depart at..." Or ask clearance delivery to call AFSS and do the same. It's better if you do it yourself because at a busy airport, clearance may be swamped. Even if you don't suspect the error, a good specialist will check the proposal list, see the error and quickly amend it. Your flight plan will appear shortly.

Airfiles are handled like any other flight plans. The only advantage to airfiling is that you don't have to listen to the Muzak while you're on hold waiting for the next available briefer.

Theoretically, if you airfile, wanting to pick up a clearance right now, as soon as AFSS enters the information, the computer will instantly address it to whatever ATC facility is involved. The flight plan should be there before you can change frequencies, assuming you're using an old KX170B.

Sometimes, should your flight plan be missing, you can ask clearance delivery to enter one for you. If it's a short route at low altitude, you may get help. A longer flight plan full of lat/longs and flight-level altitudes will most likely yield "...contact flight service on 122.65."

If your pop-up is accepted, all the controller will copy is an abbreviated flight plan—call sign, type/equipment, speed, departure point, p-time, requested altitude and route. Any mention of aircraft color, persons on board, local contacts and home base will probably be ignored so you might as well save your breath.

You may ask if airfiling anything other than a pop-up approach clearance constitutes adherence to the flight plan requirements set out in FAR 91.169 and 91.173. Or the local FSDO may ask the same question of *you.*

I have no sage advice on this one. If everything goes as planned, it's probably a moot point. Once you've landed and parked, who cares what your alternate was and who would bother to check if you got a weather briefing?

If you're really worried about it and want to cover your backside, radio FSS after you've accepted your clearance, file the route you've been assigned (including an alternate, souls on board and so on) and let the plan get tossed into the DD. That way, if anything happens, you'll have an official record of having done things in the FAR-AIM approved manner.

Finally, flight plans are living documents and like any living thing, they sometimes get killed in the strangest ways and at the most inconvenient times. If your flight plan suffers an untimely death, don't despair, refile. It may take another ten minutes, but at least you'll get to hear "Tie A Yellow Ribbon...." one more time.

Once you've received your clearance, accepted it and started on your way, things start cooking at ATC. Most of what a controller does centers on his or her radar screen. The really critical data is shown there, but there's another important tool the controller uses to keep track of who's where, and where they're headed: the flight strip.

This innocuous scrap of paper is something most pilots never see and few give any thought to, but it's vital to the controller's job. Despite the gadgetry and computers found in the ATC system, the lowly flight strip remains the primary means by which ATC tracks your airplane. Paul Berge continues his exploration of the system's insides with an explanation of how flight strips work.

Flight Strips

Anyone with more than a casual acquaintance with ATC has probably wondered: "What's all that noise I hear in the background when the controller has the mike keyed?" Beyond the usual screams, there's often a flat clicking noise, like listening to a Mah-Jongg tournament.

To lump all ATC sounds together and classify them simply as "background noise" would be too easy an explanation, plus it wouldn't fill enough space on the page. For help, we turn to the FAA's own manual: 7234.75D or Authorized Controller Sounds. This can be ordered from the Government Printing Office ($6.95).

Much of what you hear is classified as "life sounds" such as coffee spilling on keyboards, sneezes, snoring or the occasional whimper. But those plastic clicking sounds that rattle like someone's stacking Tupperware? That's the sound of flight progress strips, an admittedly unglamorous but definitely indispensable part of ATC.

Still done by paper

In the age of computers, ATC still relies on pen and paper to keep track of your IFR flight. True, you can file a flight plan through a laptop and before you even reach your airplane, FSS has sent your clearance to Brazil, all by computer. But when it comes to the actual moving of airplanes, the "turn left, turn right, cleared to..." ATC stuff, you need to talk to a human being with bad penmanship.

Your identity, your whole purpose, is reduced to a strip of paper the size of a bookmark and one that's just as easily misplaced. Originally conceived at O'Hare as a means to place off-track bets with their bookies, the flight progress strip codifies data needed to work an airplane. Approach control has one kind of strip, Center another and everyone has their own way of handling and marking strips, depending on local orders.

Okay, so you know about flight strips; but you're probably wondering what we do with all those computers built during the 1960s. Aren't they supposed to keep traffic tracked and identified? Well, yes and no. Although, the radar scope and associated computer tracking systems may be the primary source of traffic information to the controller, the flight progress strip is still the main back-up.

Should the computer fail—which it does more often than you might think—the controller dives into the strips for information. With primary radar only, that is, no discrete beacon targets, all the niceties like aircraft type, tail number and altitude disappear. Your airplane becomes nothing but a fuzzy little line on scope. With a dull pencil and updated strip-marking, the controller should be able to keep everyone separated until quitting time. When radar fails entirely—again, it does happen—a tiny paper strip is the only evidence that you exist inside the clouds. Frightening, eh?

What it all means

So what's on the strip and where does it come from? Well, there are a couple of flavors. The terminal strip is subdivided into 18 spaces, each with its on purpose. A Center strip is slightly larger and has room for 30 spaces, all with specific duties. Center controllers rely more upon their strip information than we do in approach. They're just better human beings, I guess.

An approach strip contains these basics; tail or flight number, type, proposed departure time, the route (although not always the entire route) plus (sometimes) remarks. Flight progress strips also contain information not found on the radar scope. In approach, for example, our ARTS III computer displays the target plus data block giving call sign, groundspeed, altitude and type aircraft. It does not, however, give particulars about route of flight, destination (unless within approach's airspace), departure point or remarks.

Normally, we don't care about that stuff because once we take the handoff from Center, we assume the pilot is landing somewhere, and we vector. If he's departing, we ship him to Center and he becomes their problem. But the flight progress strip tells which airport is the destination. A glance at that keeps us from vectoring you to the wrong airport almost 87 percent of the time.

If a flight plan's route is too long to fit on the strip, the computer abbreviates the flight plan, by placing *** in the route. That means the pilot has filed so many fixes, the controller must call up the flight plan in its entirety from the computer. (If you've filed one of these trying to beat a preferred route, you now know why the controller is muttering in the background.) There's also a three digit computer ID number that a controller can punch into the ARTS to request a full route on flight plan too long to fit onto a strip.

The computer may also truncate a route, if a particular sector has no need to see the entire route. The symbol ./. is inserted, indicating a truncated route. Now and then the computer throws in the ./. when it doesn't recognize a fix or understand how the pilot will navigate from one fix to the next. Keep in mind here that we're dealing with a computer with all the intelligence of a ripe zucchini.

When a route gets mangled, the controller fills in the necessary blanks. Also, it's the controller's option to assign a wrong altitude for direction of flight, if doing so is an operational advantage. That's right: the controller can assign an odd altitude for headings where even altitudes are supposed to be flown, and vice versa. When this is done, the altitude is underlined to indicate that approval request (APREQ) from the next sector is required.(That may be what the controller is doing when he claims to be on the land line; either that, or your fifth radio call finally woke him up.)

Where it comes from

As far as the air traffic control system is concerned, you don't exist until your flight strip comes squirting out of a printer a half hour before the proposed departure time. Mostly, flight plans and the strips that describe them originate in the host computer at the Center which owns the local airspace.

The strip finds its way to the right controller in various ways. Tracons have their own strip printers, as do some towers. Towers that don't have printers have to call the next facility up the line and write the information in a strip by hand.

Back in 1979, when I took my first FAA assignment at Oakland Center, it was my job to make flight strips appear. Fresh out of the Academy, I was assigned the job of "running and stuffing." Runners and stuffers are responsible for removing strips from the printers—known as FDEPs—stuffing them into little plastic holders—called "little plastic holders"—and running the strips to real controllers—

called "sir"—who expressed their gratitude by allowing us to breathe their cigarette smoke.

FDEPs resembled the old teletype machines you used to see in Flight Service; a box of clacking wheels and ribbons all delicately tuned and balanced and guaranteed to fail when the maintenance technicians had gone for the day. FDEPs have since been replaced by something better, or at least by something newer.

FDIO ("FIDO") or Flight Data Input Output has replaced the FDEP (retired FDEPs were sent to Afghan rebels). FDIO is a simpler machine, made of plastic and resembling any common outdated printer found at ComputerLand. It has few moving parts and is simple to operate, or so the manufacturers claimed when they sold the lot to the FAA, just prior to declaring bankruptcy. The FDIO prints faster, so flight progress strips can be distributed more quickly, a real plus for you if you're broiling out on the ramp waiting for a clearance.

Despite the hardware upgrade, a paper strip is still churned out for each flight, through each sector for the entire route. Sometimes, there may be dozens of strips. Every change to the flight's route or altitude, equipment suffix or ETA is noted somewhere in writing. On Center strips any change to route, altitude or other information that must be passed along to the next controller is circled in red. This indicates that coordination has been accomplished.

The information may also be entered into the computer, but the red circle is a visual reminder that everything's taken care of. In approach, we don't use the red pencils. We simply place a check mark beside coordinated information which means it's sometimes difficult to tell what's been coordinated when a pilot's made a string of requests. Ever ask for six different routes and nine different altitudes, only to have the handoff wind up in the ozone? Beginning to get the picture here?

Paper reality

In a radar facility, the computer does most of the grunt work in tracking airplanes. But at nonradar facilities—there are still quite a few—the arrangement of flight progress strips produces a mental image of traffic. Altitudes are recorded, fix passing times are noted. When a pilot leaves an altitude or reports level at another, the information is recorded, so the controller knows what altitudes are still available. When a pilot reports passing a fix, the time is recorded, as well as the estimate for the next fix.

All this information forms the three dimensional picture that, hopefully, matches reality. Without radar, the controller can't actually see the traffic, so in the event of possible conflicts, a "paper stop" might be issued to one pilot: "Cessna 123, cleared to (an alternate fix) and hold...expect further clearance at..."

Despite what you may read about recurrent training, radar controllers are not enthused about reverting to non-radar procedures. It takes practice to be able to read a stack of strips and understand the traffic picture, especially if you're used to picking it off the radar datablock.

At some facilities—particularly busy ones—one strip is used and notations are made to it to record additional operations, such as practice approaches. At Des Moines, we write a new strip for each approach, a practice that compounds workload without adding to air safety.

Again, keep this in mind if you're making a lot of requests. Could be, the controller just doesn't have time to write it all down. A neophyte controller, wrongly assuming strip marking is paramount, can spend an inordinate amount of time carefully penning strips and never watching the scope; similar to a new pilot scanning instruments while never looking out the window. Experience convinces the controller to make quick notes and backfill the strip marking later.

There's supposed to be a standard shorthand strip markup but, in fact, it varies a little from Center to approach and from facility to facility. These notations are both for the controller working the traffic and for anyone else who might need to pick information off the strip.

One thing we're often asked about is remarks. Does it do any good to add "Training Flight" or "Practice Approaches" to the remarks field? Maybe, maybe not.

Most controllers don't normally read the remarks section so your request may go unnoticed. I recommend this: When you check on the frequency, make your request specifically with the controller.

They don't fade away

If you've ever been in an ATC facility, you've probably noticed that once the aircraft disappears over the horizon or parks on the ramp, the strip isn't simply tossed in the trash. That's because a flight progress strip is considered a legal document. Daily, the strips are counted, wrapped in rubber bands, blessed and stored for use in case of an accident or incident investigation.

After several weeks, if no one's died, disappeared or fallen into FSDO's hands, the strips are discarded. Should an investigation arise during that time, all strips pertaining to the flight will be included in the investigation package, along with voice tapes. Trust me, being confronted with a strip in your own chicken scratch can be embarrassing, especially when a lot of unsympathetic strangers are asking, "...don't you follow the 7110.65 strip marking rules?!" Don't spread this around, but strips have been known to be dummied up after an incident.

You may think that with new ATC computers just around the corner, the flight strip's days are numbered. Don't bet on it. Ever since I became a controller, I've been hearing the prophesy that flight progress strips would soon be replaced with video strips. It's been 13 years. Someday they may be outlawed, and no one will grieve their passing. You should, however, be aware that controllers can only think as fast as they can write, and until something replaces the pencil, you may be asked to repeat yourself.

"Cessna 1234, contact New York now on 120.55. Good day."

That simple frequency change is familiar to all instrument pilots. It's a lot more complicated than it seems on the surface, though. Veteran Chicago tracon controller Scott Hartwig gives a behind-the-scenes look at how controllers coordinate with one another to make handoffs as smooth and seamless as possible.

What Happens at the Handoff

The Pilot/Controller glossary defines a handoff as "an action taken to transfer the radar identification of an aircraft from one controller to another if the aircraft will enter the receiving controller's airspace and communications will be transferred." That's kind of a mouthful for what, to the pilot, amounts to a simple frequency change.

From the other end of the radar scope, however, the instruction to change frequencies actually culminates a process that may take several minutes. As a pilot, you only hear one side of a number of conversations that may occur as your airplane is passed from controller to controller. If all goes as planned, the handoff is seamless and efficient. But like everything else in IFR flying, there are occasional snags. Knowing a bit of how handoffs work will give you a better sense of what's going on.

Auto handoffs

When you're planning and flying IFR routes, it helps to remember that air traffic control is made up of many sectors—both horizontal and vertical. Flying across several sectors is like walking through a residential neighborhood. In order to cut across someone's yard, you really ought to have permission, and you're expected to abide by the owner's request to stay out of the flower bed. In the ATC analogy, each controller "owns" a particular yard and is responsible for who uses it and how.

On the radar display, the controller can see not only his or her own area but portions of the adjacent airspace. Aircraft appear as symbols accompanied by data blocks containing two or three short lines of information about the flight, including the call sign, altitude and speed and, depending on the type of facility, they may show a computer identification number. On tracon scopes, the aircraft target appears as a small symbol, such as a letter or a triangle. On ARTCC scopes, which depict digital representations of target positions rather than the actual target position, the airplane appears as a slash, a dot or a plus sign, depending on what kind of information the tracking computer has about the airplane.

The data blocks automatically follow the aircraft to help the controller keep track of its identity. When an airplane is about to leave one sector and enter an adjacent one, it must be identified and accepted by the receiving controller. These days, that transfer is almost always done automatically and it's called (surprise) an automated handoff.

Here's how the automated handoff works: First, the transferring controller starts the process by using a trackball (kind of like an inverted computer mouse) to move a cursor on his display to the target to be handed off. Then, by pressing a button, he makes the datablock flash on the receiving controller's display. At this point, the aircraft is not yet in the receiving controller's airspace but he can see portions of adjacent areas and thus will see the oncoming flashing datablock.

To accept the handoff, the receiving controller moves his own cursor to the aircraft and pushes a button that causes the datablock to cease flashing. At the same time, the datablock on the sending controller's display will flash momentarily to acknowledge acceptance of the handoff. In case the sending controller misses the signal, there's also a little symbol on the block to indicate that the handoff was properly accepted.

In some cases, the handoff is initiated automatically (there's that word again) by the tracking computers involved, but it still requires the receiving controller to move his cursor or "slew" to the target to accept it. Unfortunately, some facilities aren't as automated as others. Some may not have any sort of tracking computers to supplement their radar, while others have a computer that doesn't interface completely with the rest of the system. In these cases, the handoff has to be done manually via a telephone voice line set up just for this purpose.

Manual handoffs require the sending controller to verbally describe the position of the airplane to the receiving controller. That conversation might sound something like this:

Controller A: "Rockford, Sector Three, handoff."

Controller B: "Rockford."

Controller A: "Five miles west of Dupage VOR, Cessna One Seven

Zero Lima Bravo, level 4000."

Controller B: "Radar contact, B-D"

Controller A: "S-T"

Controllers usually close out conversations with their operating initials so that, should the need arise, they can be identified later. Remember, all of these conversations are tape recorded and kept for at least 15 days, or longer if needed. By the way, I use ST because an old-timer at ORD was already using SH.

Down the line

The handoff shouldn't be a complete surprise to the receiving controller. He will usually (but not always) have, in a bay near his display, a printed flight progress strip containing flight plan information such as call sign, type aircraft, route and altitude. The controllers then verbally verify the position and any other pertinent information such as assigned headings or airspeed restrictions.

Manual handoffs don't usually involve a lot of time but they do require the controller to leave the frequency. When he's on the voice line, aircraft communications are usually routed through a loudspeaker at his position. Depending on the equipment, he may or may not be able to talk to aircraft while on the phone but he may still hear calls over the speaker. Hence, the often-heard "Aircraft calling Chicago approach say again, I was on the land-line."

A manual handoff is just one reason a controller might be off the frequency. Some of the busiest sectors will employ a second controller to do the phone work and to assist the radar controller. But very often, controllers work multiple frequencies and sectors without an assistant.

A variation on the voice-line method is for the sending controller to physically point to the aircraft on the receiving controller's display. In some cases the two controllers may be seated right next to each; they may even share the same display. In this case, the sending controller merely points to the aircraft on the receiving controller's display.

Since each controller is responsible for what happens in his airspace, the sending controller has to complete the handoff before the traffic enters the next controller's area. In other words, he needs an official okay from the next guy. He can't simply let his traffic go barging into the next guy's "grass" (old controller slang for airspace). To do so would likely cause an operational error or in controller slang, a "deal."

Similarly, the receiving controller can't turn, climb or descend the traffic until it reaches his airspace, unless the sending controller concurs. This is why a pilot's request for altitude or route changes near the handoff point is often delayed. The coordination required between two busy controllers might take longer to accomplish than the short delay in waiting until the traffic crosses the boundary.

Sometimes the handoff procedure is enhanced by a Letter of Agreement between facilities. In fact, it's a must for an automated handoff. An example of a letter between Chicago approach and Rockford approach might say that traffic from Chicago, traveling through Rockford's airspace, be established on a 270-degree heading, at 6000 or 8000 feet, the final choice of altitude resting with the Chicago controller. Rockford would protect for either case, and wait until the traffic was in their airspace before making any changes.

This can also be done on an informal basis between controllers. For example, because of strong south winds on a given day, controller A coordinates with controller B to leave all traffic on 260-degree headings instead of 270, as per the letter.

When things go sour

Sometimes things don't go as planned. There are a number of reasons that a handoff might not be completed soon enough, such as interruption in phone services, a breakdown of the interface between computers or the receiving controller being too busy to accept more traffic.

He may be too busy to even talk to the sending controller, in which case the sending controller would initiate some sort of holding procedure before encroaching on the next guy's area. The holding may be a bona-fide holding pattern or simply some delaying vectors or 360s (what a controller means when he says "spins") until something can be ironed out with the next guy.

Most of the tracking systems used by approach controls do not have the ability to "talk" directly to other approach controls. The individual approach control systems are linked through the larger computers in the ARTCCs that overlie them.

In fact, at one time, Springfield approach, which is located under the boundary between Chicago and Kansas City Centers, could only make automated handoffs to Kansas City. The bulk of their traffic, however, traveled to Chicago Center, thus requiring manual handoffs. Even though Springfield's airspace adjoins Peoria approach, they could not have automated handoffs between each other because they "belonged" to different Centers.

Why pop-ups don't work

Pop-up IFR plans sometimes pose a problem as a result of this limited capability between terminal radar systems. Within a given approach control airspace, handoffs between controllers are fairly simple, but to automatically hand off between approach controls requires that a flight plan be entered into the Center's computer before information can be forwarded to the adjacent approach control.

This means that if you depart Dupage, for example, and ask approach control for an IFR to Waukegan, you'll probably get it with little delay because both are in Chicago's airspace. But if you ask for the IFR to Kenosha, which is only 8 miles farther than Waukegan, you'll likely be greeted with less enthusiasm.

Kenosha is in Milwaukee approach's airspace; thus, all of the flight plan information would have to be forwarded manually, or a flight plan would have to be entered into the Center's computer in order to automate the handoff to Milwaukee.

The same is true for VFR advisories: To go from sector to sector within a facility is relatively easy. However, when transiting from one facility's airspace to another, such as from Chicago Center to Chicago approach, a manual handoff is required. More often than not, you'll hear "radar service terminated, contact Chicago approach on 126.8."

The receiving controller may not be too busy to handle you, but he may be too busy to spend the time messing with a manual handoff. In other words, it's easier for him to reidentify you with a new transponder code in his system than it is to find you via the manual process.

Unfortunately, the individual sector and facility boundaries aren't published and, in many cases, it would be difficult to do so because they sometimes change as often as the weather. At Chicago, as is the case at many locales, the shape of the airspace sectors is dictated by the runways in use at O'Hare. Therefore, it's difficult for pilots to anticipate when a handoff will take place unless they're on a familiar route.

It's also not always appropriate to assume that because the frequency doesn't sound busy that the controller isn't. Some sectors require a lot more manual coordination than others. For example, an arrival controller at ORD rarely spends any time on a voice line (he's got a helper), whereas an arrival controller for Dupage does a significant amount of coordination with receiving towers at Aurora, Dupage, Pal-Waukee, Glenview and Waukegan. It's not unusual for the Dupage controller to spend more time on the voice line than on the frequency.

So, next time you ask for a pop-up IFR clearance or call an approach control and get no answer for a few minutes, you'll know why. There's a lot going on behind the scenes that the pilot never hears.

• Section Two •

Unraveling Clearances

Eliminating Clearance Confusion

Clearances are part and parcel of being an instrument pilot. They're also behind many violations, misunderstandings, and even accidents. There's simply no room for misunderstanding here: without a well-understood clearance the pilot has no firm idea of what ATC wants him to do, and the controller doesn't know where the pilot is likely to go.

IFR *editor and charter pilot Paul Bertorelli here passes on some hard-won tips for ensuring that there's no misunderstanding when it comes to getting a clearance.*

Make Certain Everything's Understood

As an avocation, flying seems to attract hard-driving, take charge characters who tend to seek rather than weasel out of responsibility. Good thing, too, because aviation is culturally biased to saddle the pilot with so much responsibility that he or she gets blamed for everything from shock cooling (power changes too abrupt) to altitude busts (the tape clearly says *4000* feet).

To a certain extent, even if ATC screws up, there's a specific provision in the FARs to lay it off on the pilot. It's FAR 91.75, which has to do with following clearances. To paraphrase slightly, this passage says that a pilot has either to follow a clearance, seek an amendment or declare an emergency, in that order. In the case of a clearance that's confusing or just plain wrong, it's the pilot's responsibility to ask ATC for clarification.

That last requirement escapes many of us. Simply staying current on procedures is hard enough. Who has time to analyze every clearance for correctness before flying it? When in doubt—and that's often enough—most of us fly on faith, assuming that the voice of authority (ATC) knows its stuff.

In reality, however, controllers probably err as often as pilots and for some of the same reasons. But because pilots know relatively little about proper ATC procedures, the minor errors go unnoticed.

Bogus clearances

Between them, the AIM and the 7110.65 (Air Traffic Control) devote something like 200 pages to IFR operations, including specific examples of how clearances are supposed to be issued, right down to the exact phraseology. Keeping all of this in mind while you're trying to nail the localizer in turbulence or sequence five airplanes for the approach is not easy. If you don't know the stuff cold, almost as second nature, it's probably better not to think about it at all for fear of making things worse through lack of concentration.

As we'll see in the next chapter, controllers routinely issue clearances with inadvertent gaps that make them practically unflyable. Most of us accept these clearances, betting on the good odds that radar vectors will eventually iron things out. Nervous (or maybe cautious) pilots pump the controller for something more specific, which is probably not a bad idea. But if you do it consistently, you'll spend a lot of time on the frequency tidying up loose ends.

Here's an inconsequential example: Last summer, on a checkride, we were shooting practice ILSs to runway 13 at Atlantic City. By coincidence, two other airplanes were doing the same. It was a slow afternoon and the controller had no traffic other than the three training flights.

Since we were closest, we were first in line with a vector to final and this clearance: "Nine Three Uniform, you're three miles from NAADA...maintain 1600 until crossing NAADA, you're cleared for the approach."

The clearance got us onto the approach all right, but it's wrong. NAADA is a LOM and thus it's the non-precision FAF. Maintaining 1600 until the FAF is fine for the LOC, but it would have put us 88 feet above the glideslope before descent began. That's only a dot or or so of deflection. No big deal. But controllers aren't supposed to issue altitude restrictions that result in GS intercept from above.

Technically, I suppose, we should have called the controller's attention to the error and asked for a corrected clearance. But who wants to tie up the frequency over an 88-foot discrepancy? We just intercepted the glideslope and flew the approach. I assume the other aircraft did the same because the controller issued the same incorrect clearance five or six times.

Sometimes it matters

You can't always be that blasé, though. Sometimes accepting a bogus or misunderstood clearance can mean big trouble. Example 2: Another slow afternoon, inbound to home base at Oxford, which has an ILS. Twenty miles southwest of Bridgeport VOR (an IAF for the ILS) at 3000 feet, we asked the controller for direct Bridgeport followed by the transition and approach clearance for the ILS. The weather was marginal VFR in moderate rain.

The controller immediately came back with this : "Yeah, okay, go ahead and do that." Thinking I'd understood his intent, we crossed Bridgeport and started a descent to 2300 feet, the transition altitude. Thirty seconds later, the controller noticed our Mode-C trace and asked if we were descending.

"One Three Hotel, affirmative, we're out of 3000 for 2300 feet. We're flying the transition," I explained.

"I didn't clear you for the approach, maintain 3000," the controller replied, explaining that he had conflicting traffic ahead. Evidently, no separation bust occurred but I slipped an ASRS form into the mail anyway.

The clearance was definitely improper but the controller had an accomplice. I'd asked for the transition and cleared myself for the approach, hoping the controller would rubber stamp my plan. When he said "yeah, okay," I mistakenly assumed that he had done just that. I should have queried and gotten the correct clearance, which would have been something like: "Cleared direct to Bridgeport, cross the VOR at or above 3000, cleared for the approach."

Everyday errors

According to the phone calls and mail we receive from pilots who fly a lot of IFR, this sort of thing happens every day. NASA's ASRS program takes in between 2000 and 3000 reports a month; many deal with bogus or misunderstood clearances, quite a few of which are submitted by controllers who detect their own mistakes or have them pointed out by others.

Some incorrect or improper clearances are simple errors due to fatigue or miscommunication; the controller knows the correct procedure but just doesn't communicate it clearly. Other errors—such as the approach clearance we got into Atlantic City—are institutional shortcomings related to poor training or lack of currency, the same thing that gets pilots into jams. Either kind of error shouldn't be a problem for the pilot who's a serious student of the IFR system. He or she will know when it's safe to ignore a minor error and when a query is necessary to straighten things out.

Books aren't the same

Apart from miscommunication and lack of currency, some errors stem from the fact that pilots and controllers often have no shared understanding of how a particular procedure is supposed to work. The AIM and 7110.65 can confuse rather than illuminate the issue. The AIM is written for pilots, the .65 for controllers and in addition to the fact that the two documents don't have parallel language, how many controllers or pilots are fluent in both books? Just try to find a current AIM in an ATC radar facility.

A classic example is the continuing confusion over VFR-on-top clearances and VFR climbs and descents. A Florida reader recently phoned to explain that during an Operation Raincheck visit to the Orlando tracon, he'd asked about this very topic. The controllers giving the tour explained that they hadn't approved VFR-on-top because they hadn't heard of it; referring to the .65 evidently didn't clarify the intent of the procedure.

Comparing the language in the two books, it's easy to see why a controller wouldn't be up on VFR-on-top. The AIM explains why a pilot would request an on-top clearance, "this would permit the pilot to select an altitude of...his choice." The .65, on the other hand, merely says the controller may issue VFR-on-top if the pilot requests it, along with a short list of restrictions. No other explanation is given. To a controller who's not a pilot—and that's most controllers—it may not be at all obvious why a pilot would make such a request.

VFR-on-top and VFR climbs and descents represent a vastly underused means of minimizing delay and/or simplifying routing. But because some controllers turn down such requests because they're fuzzy on procedures, pilots stop asking and the potential is lost.

Contact approaches are another example of an underused (and misunderstood) procedure that slips into the cracks between the pilot's and the controller's knowledge. It does little good if the pilot requests a contact, only to have the controller deny it because he or she isn't sure of what's required. An exchange we heard on the approach frequency into Westchester County last fall illustrates what can happen when a controller is confronted with an out-of-the-ordinary (but perfectly proper) request.

The aircraft was inbound to Westchester from the east, being vectored for the ILS 16. Reported weather was 1200-foot broken, 2-1/2 miles in fog, although the vis looked much better than that as we departed 16 toward Long Island. The reported weather was too poor for a visual and apparently hoping to avoid out-of-the-way vectors to join the ILS, the pilot asked for a contact, a request that evidently surprised the controller.

"Six Four Quebec, you want a contact approach? Okay...uh, standby," came the reply. A moment later, the controller had a request of his own: "Six Four Quebec, can you accept a visual...no, uh, standby." I fought the urge to break onto the frequency and say "Hey look, he's got a mile and clear of clouds and ground vis is 2-1/2 miles. What's the problem?"

Finally, the controller rejected the request and issued a heading for vectors to the ILS. Thirty seconds later, the controller asked the pilot what his flight visibility was and upon learning that it was three miles, he cleared the pilot for a contact approach.

There may have been extenuating circumstances involving other aircraft but clearly, the controller wasn't sure of the requirements for a contact approach and just to be safe, he turned down the request. While the pilot was standing by, I suspect the controller was polling the radar room for some advice on issuing the proper clearance.

What to do

As we said, confusing and/or bogus clearances are an everyday fact of life in IFR flying and the FARs leave no doubt about who's supposed to iron them out; it's the pilot's responsibility. In order to know a bum clearance from a good one, you've got to keep current on both IFR flight procedures and, to a certain extent, ATC procedures.

Short of enrolling in the ATC Academy at Oklahoma City, the best way of doing this is to regularly review both the AIM and the 7110.65. You can get a copy of the AIM at just about any FBO and even some book stores; the 7110.65 is available by subscription from the Government Printing Office (Washington, D.C. 20402, 202-783-3238).

Sure, both are weighty documents and dry reading at that but the sections that deal with clearances, IFR procedures and routing and approach procedures are short enough to review in less than a half hour. If you're really ambitious, copy the relevant portions of 7110.65 and file them in your Jepp binder or flight bag. That way you'll have something to do next time you're weathered in or the passengers are late.

The Unflyable Clearance

It doesn't happen often, but every once in a while ATC will give a pilot a clearance that simply does not make sense. It may be unflyable for some reason, or the controller may make a mistake in reading it—for example, using the wrong name for a recently changed navaid out of habit—or perhaps there are missing instructions that would leave the pilot out on a limb in the event of comm failure.

Naturally, in the FAA tradition of making sure that pilots are responsible for everything, it's up to the pilot to make sure that the clearance is proper before accepting it. All too often, this does not happen: we usually just read back whatever the controller says without thinking about it first.

There are other occasions, though, when a clearance doesn't seem to make sense when it really is a legal, flyable one. This is never a desirable state of affairs, however. Whenever a pilot's status is in doubt, either on the part of the pilot or on the part of ATC, the situation becomes ripe for misunderstanding and potential disaster. IFR *contributor and Cessna pilot Brian Weiss relates the tale of a clearance that left both him and ATC in doubt as to whether or not he was actually on an IFR flight.*

A Little Bit IFR?

The laws of nature being soundly in place, it's still impossible to be a little bit pregnant. It is, however, entirely possible to be a little bit IFR. Or so it would seem.

I discovered this a few years ago when, at approximately 1515 local, I departed John Wayne airport on an ARSA clearance, westbound for Santa Monica. On departure I found the three miles of visibility was indeed a marginal three. Suspecting that things would go from bad to worse to the west, I asked departure to convert my clearance to an IFR, tower-to-tower. They told me to call Coast with my request.

"Hello Coast, Cessna 5389K, on an ARSA clearance, requesting tower-to-tower IFR to SMO."

"Yes sir, remain VFR at 3500 feet while we work on that clearance."

"89K, clearance."

"Yessir, fire when ready."

"89K is cleared to Santa Monica Airport, depart SLI on the 270 degree radial for radar vectors LAX, LAX 316 radial, maintain VFR at 4000."

Hmm. Not the usual clearance. But it had been an unusual day that included the President in and out of SMO, a 30-minute wait for an IFR departure out of SMO, and clearance delivery at SNA going off line for ten minutes. Anything was possible.

It was not the first time I'd been issued a hybrid here's-your-IFR-clearance, remain-in-VFR-conditions clearance, though the mix of an IFR altitude and an IFR clearance with instructions to remain VFR was an invitation to confusion.

I accepted the invitation. Five minutes and one controller later, Coast calls with an amended clearance. Great, they'll now clear up all this confusion.

"Depart SLI on the SLI 270 radial, radar vectors LAX, LAX 046 radial to join V186, V186 DARTS, direct." No further news on altitude, no VFR restriction. Fine, that's the standard clearance from that area to SMO. I'm IFR, and free at last.

Not quite. Coast hands me to Los Angeles approach, which hands me to another LA controller who asks for an accelerated descent to 3000 feet for the turn north up the shoreline and past LAX. I tell this person I'll be happy to take a visual into SMO. He says unable due to traffic, you're number four for the approach, be so kind as to fly 030 degrees off LAX. And contact LA controller number three.

"Hello Los Angeles, Cessna 5389K, level at three, 030 vector."

"89K, are you a VFR flight?"

Pardon me sir? I'm the owner of not one but two IFR clearances, several IFR altitude assignments, and one turndown for a visual approach. And now you want to know if I'm a VFR flight? You folks getting sufficient ventilation in the little room where you live?

Who, he demands rather curtly, gave you an IFR clearance? Perhaps Cessna 172s aren't allowed to have them any more, I think, or maybe we've gone on an odd-even allocation system and it isn't my day?

I inform LA that I have a perfectly healthy IFR clearance from the friendly folks at Coast. This is-it-fish-or-fowl flight eventually terminates in an IFR approach to Santa Monica.

Curious beyond all belief, I call my friendly local Accident Prevention Counselor, John Plueger. I'm a big believer in the Two Heads Are Better Than One theory of solving ATC mysteries. What, I ask him after describing the sequence of events, is wrong with this picture?

We agree that (1) I probably should have pressed the case with the original controller as to whether the initial clearance was IFR, (2) when the amended clearance came, with no stated restriction to VFR conditions, it was an IFR flight, (3) there should have been no reason for LA approach to be confused about my status, and (4) Coast had some explaining to do.

Next stop, Coast approach, via the land line. It's now about 1645 local. A very courteous supervisor answers the phone and listens to my tale of woe. He says it's interesting that I should call about that because the business of issuing IFR clearances with instructions to remain VFR has been bothering him a bit, too.

He then spends considerable time carefully and completely explaining to me how things really work at Coast. "We issue IFR clearances with instructions to remain VFR", he says, "to expedite things. Sometimes we can make the system move better for you by issuing a clearance but keeping you temporarily VFR with the current controller."

"That gives us and the next controller more flexibility in working you into the system. Otherwise, you would experience more delays." We both agree, however, that when I was read the amended clearance, sans a VFR restriction, that I was IFR, or at least had a right to conclude that I was. Why LA was in doubt about my status remains a mystery, since if I was VFR he wouldn't have had a strip on me.

We also both agree that since nobody is specifically delegated to tell the pilot when he is definitely hard IFR, that this bootleg procedure is ripe for confusion of the sort that often leads to incidents of the most unfortunate kind. My feeling is that either this procedure should be given official sanction and specific terminology, or it should be stricken so that like pregnancy, IFR is something you either are or aren't.

Sometimes a clearance isn't flyable, though it seems to be so on the surface. It may not make any difference in the real world, but the system is predicated on a pilot always knowing how to get from where he or she is at any moment to the destination. Gaps in this knowledge mean that a pilot could conceivably get into a situation that can't be gotten out of.

In this section IFR *editor Paul Bertorelli explores the ramifications of unflyable clearances, and how they might affect an instrument pilot.*

How Are You Navigating?

Given the amount of IFR traffic that flickers across their radar displays on an average day, controllers in the northeast corridor (and elsewhere) are remarkably resourceful at working out pilot-requested reroutes. Most controllers, no matter how disgusted they are with the FAA, consider it a point of professional pride to know the system well enough to manipulate it in everyone's favor, sometimes in the face of illogical local procedures.

Despite having to work around rigid procedures themselves, controllers tend to be perplexed (if not amused) when a pilot becomes nervous because his clearance sounds a little *too* flexible. Things can go to pieces quickly when a pilot starts to wonder how the controller expects him to get from the airport to a fix or vice versa.

I was reminded of that a while ago as I completed a short ferry flight from Groton, Conn. to our homebase, at Waterbury-Oxford. The weather was perfect VFR but I always file at night so I picked up a clearance from Groton tower just before it closed. The controller read off the usual preferred route, a clearance I'd heard so many times I didn't bother to copy it: "Right turn after departure, V374 CREAM, direct, maintain 3000, expect 6000 ten minutes after departure."

Leaving Groton, I'd been the only aircraft on frequency and even after the handoff to New York, things remained unusually quiet. So quiet, I guess, that the controller decided to play a little Pull the Wings Off the Fly.

New York: "Ah, 46 Yankee, how'd you plan to fly this clearance?"

Me: "Huh?"

New York: "Well, you got CREAM then direct Oxford, right?"

Me: "That's affirm."

New York: "So how you gonna get from CREAM to Oxford."

Okay, I thought, I'll bite. I peered at the chart in search of an answer. Trouble is, CREAM is in the middle of Long Island Sound southeast of New Haven. It's formed by seven radials, none of which come closer than about ten miles of Oxford. You can get to CREAM, but you can't really go anywhere from there. Aircraft departing CREAM are usually given radar vectors to their destinations so that's what I suggested.

New York: "Okay, but what about lost comm?"

Me: "Then I'd go direct CLERA and shoot the ILS."

Even before I'd uttered the last of that sentence, however, I knew he had me. CLERA, the LOM for Oxford's ILS had been notam'd out of service for several weeks. I knew about the notam and evidently, so did the controller. But both of us also knew I'd accepted what appeared to be an unflyable clearance anyway.

Actually, upon further consideration, the logical response to lost comm would have been to fly direct to Bridgeport (an IAF for Oxford's ILS) and then complete the approach. But if that's so obvious, why didn't Bridgeport appear on the clearance in the first place, for lost comm purposes? Then there would have been no doubt.

It turns out that the CREAM routing is a preferred departure arrival route for several airports. The Oxford leg was intentionally left off for the simple expedient of a shorter clearance and to save computer memory.

Happens all the time

My Oxford trip was a relatively benign example of a common occurrence in IFR flying: Whether by accident or design, many clearances simply don't cover all the bases, or at least they don't appear to. In some cases, the discontinuities are so subtle or trivial that you don't notice them until, as in my case, someone points them out. Under other circumstances, a clearance can have such a gaping hole that it creates overwhelming confusion. In that case, pressing on in the blind hope that ATC will look out for you is a mistake. The burden of finding the true path rests squarely on the pilot's shoulders.

A handful of fatals (and quite a few violations) have occurred as a result of muddy clearances. The most spectacular fatal happened in 1974 when a TWA crew misunderstood an approach clearance, misread the plate and descended into terrain near Dulles International, killing all aboard. Another fatal occurred several years ago when a Cessna 210 pilot, the recipient of several confusing transmissions from ATC, obviously misunderstood his responsibility to assure his own obstacle protection. Along with three passengers, the pilot was killed when the 210 slammed into high terrain west of El Paso.

Most unflyable clearances are more mundane than threatening, but they still cause confusion. The worst discontinuities seem to involve departures, often from uncontrolled airports but from major terminals as well, where SIDs are routinely issued to departures. Typical is the clearance that doesn't provide a heading after takeoff or merely instructs the pilot to intercept an airway without specifying a particular fix.

That's exactly what happened to the pilot in the El Paso accident. He was instructed to intercept an airway but given no particular fix to fly to. The pilot evidently expected radar vectors to the airway and the controller evidently thought the pilot was navigating to the airway on his own. Both pilot and controller dropped the ball but the pilot paid the price.

Legal ramifications

The price of an unflyable clearance is, fortunately, rarely an accident. Usually, it's a violation, and the pilot is likely to take the heat.

As FARs go, 91.75 is relatively clear on the issue. It says, in part, that "if a pilot is uncertain of the meaning of an ATC clearance, he shall immediately request clarification from ATC." That's easy enough to do while you're reviewing your route before takeoff but not so easy when the amendment comes, machine-gun-like, during climbout into low IMC.

The temptation is to just accept the clearance, assume ATC knows what it's doing and concentrate on the flying. But as the crew of a California commuter found out, ATC makes its share of mistakes, some of which may be sorted out in court, at the pilot's expense.

This particular crew had been cleared from San Francisco to Stockton along V-244S. The airway had two components. The first was Oakland's R-093 to SUNOL intersection. At SUNOL, the airway jogged left 44 degrees and proceeded on the R-229 to Stockton. On the ground, the crew was cleared to fly direct to SUNOL. But after takeoff, the controller issued this amendment: "Continue left, heading 060, join V-244S, resume normal navigation."

The pilots understood the clearance to mean they were to maintain a heading of 060 until intercepting the R-229 to Stockton. Unfortunately, the controller "meant" for the crew to intercept the R-093, thence to SUNOL. As a result, the FAA went after the captain's ATP ticket for deviating from a clearance (91.75) and for the ever-popular FAA cudgel, 91.9, careless and reckless operation of an aircraft.

Naturally, the captain appealed to the NTSB and after a two-day trial, the judge threw out the FAA suspension, ruling that the captain acted reasonably and had interpretated the clearance just as any experienced pilot would. The judge faulted ATC for using imprecise phraseology. The NTSB later upheld the judge.

The controller's handbook (FAA 7110.65) describes, among other things, specific phraseology for clearances. That's why controllers use the same language, no matter where they happen to be. In defending himself against a clearance bust, a pilot may argue (and be able to prove) that a controller used non-standard phraseology.

One bizarre response to this has been the FAA arguing that an amended clearance is really "an instruction or a directive." Technically, there is a distinction but it's one with precious little difference. A clearance is "an authorization by ATC for the purpose of preventing collision between two known aircraft, for an aircraft to proceed under specified conditions." A directive, on the other hand, requires the pilot to "take specific action, i.e. turn left, heading two five zero, go around," and so forth. Unfortunately, the argument is pretzel logic. If the clearance from which the the pilot deviated was merely an "instruction," how can the FAA prove that a clearance was violated?

Another case illustrates how a careful readback instead of a brisk "roger" might help avoid a violation. A crew was violated for taxiing onto an active runway contrary to ATC instructions. The FAA played tower tapes that clearly showed that the crew was instructed to hold short. Rather than reading back the instruction, however, the crew responded with a roger.

In their defense, the pilots claimed they didn't hear the full instruction because of a keyed mike. But the NTSB upheld the FAA's suspension of both pilot's certificates. The lesson to be learned here is succinctly summed up in NTSB-speak: "A pilot who elects to acknowledge a clearance without a readback cannot be heard to complain that he did not understand the transmission or did not hear it at all."

The most onerous thing about defective clearances is how easily a pilot can be lured into one. If the controller truly erred, defending against certificate action is not difficult. But it's not cheap, either. It's best to avoid the bust in the first place by carefully reviewing the clearance on the ground, before it's accepted.

Remember, as far as the FAA is concerned, the burden of proof is on the pilot, not the agency. You're supposed to have "all available information" before departing on a flight and these days, that can be nearly impossible. Failing that, a checklist of the basics is vital. Ask yourself a few questions before departing.

Is the clearance legal? Do you have onboard the operating equipment to fly the route as cleared? Have you checked notams to see that all the required navaids are on the air? It's not at all uncommon to be given a clearance based on navaids that are down for service or unusable. One pilot I know was cleared to a final approach fix that was deleted from the charts in1969. The controller had simply named the fix from habit.

Question any clearance that doesn't appear to be flyable or legal. Once airborne, read back in full any amendment to a clearance and, for the tape, ask the controller to clarify anything you don't understand.

It's the pilot's call

Between them, the FARs, the AIM and the controller's handbook go into exhaustive detail about clearances, even to the extent of providing examples and phraseology. Not much is said about what constitutes a flyable clearance, however. You won't find a nice crisp paragraph that summarizes exactly what a legal clearance is. And unless they happen to also be pilots, controllers don't often sense when, because of a route break or a phantom fix, a clearance appears to be unflyable in the event of lost comm.

But it's the pilot's responsibility to get a confusing clearance clarified if he or she is uncertain about how to proceed or worried about what to do if the radios quit.

In the case of departures, such as the 210 leaving El Paso, the AIM is quite clear on who's responsible for assuring a correct and safe departure.

The pilot must determine what action will assure a safe departure; ATC doesn't assume obstacle clearance until the controller begins to issue radar vectors, not just when he or she confirms radar contact.

The same thing is true if a pilot accepts a vector or a pilot nav SID. Some SIDs are so simple that you can't go wrong: "All runways, fly runway heading to 2000, thence..." Others may be amended to suit the traffic flow on a particular day and this sometimes introduces route gaps. If you're comfortable navigating the SID as given, proceed. However, if there's any doubt, ask the controller for clarification. It's your responsibility to accept only a clearance that you can fly.

Cutting corners

Whether they are asked to or not, some controllers take it upon themselves to shave a few miles off an IFR trip. This is especially true in the northeast corridor, where preferred routes seem to go everywhere but to the destination. Corner cutting, however, occasionally introduces some ambiguity into a clearance.

A commonly heard phrase is "when able, proceed direct ABC" or "when receiving ABC suitable for navigation, proceed direct." The "when able" part confuses some pilots because it introduces pilot discretion into what had been a cut-and-dried clearance or route. In the departure context, "when able" means that you can proceed as cleared as soon as (a) you have assured your own obstacle clearance and (b) you have some means of getting to where you've been cleared.

The obstacle clearance is self-explanatory; the means of getting there is more subtle. Since IFR navigation is generally done with ground-based navaids, the pilot is supposed to know under what circumstances those navaids can be used and, in theory, shouldn't accept a clearance that requires using a navaid beyond its standard service volume. Standard service volumes are given in the AIM, along with the explanation that they apply only when operating on random or unpublished routes.

So, if you're given a clearance to proceed direct "when able," you're supposed to know if you're within the required service volume. The controller's handbook has the very same table of service volume limitations that the AIM does but controllers aren't always sensitive to navaid range considerations. Normally, they don't really have to be. Controllers are allowed to approve off-airways routes beyond standard service volumes, so long as the aircraft is radar monitored, which is most of the time. That's why it's perfectly acceptable to be cleared to intercept a localizer 25 miles from the airport, seven miles outside the normal service volume.

Blundering around

Of course, all this assumes that you're navigating electronically. However, nothing says that you have to navigate by VOR or NDB or by IFR-approved loran or RNAV. All the FARs say is that you have to fly the clearance as accepted and, in controlled airspace, fly the centerline of the airways or, if off-route, along a direct course between the fixes defining the route. FAR 91.205 requires navigation equipment appropriate to the ground facilities to be used but it doesn't say you *have* to use ground facilities.

Radical as it sounds, dead reckoning while IFR is allowed. Although most CFIIs (myself included) don't spend much time doing it, FAR 61.65 does require instrument candidates to be schooled in dead reckoning skills "appropriate to IFR navigation." I take that to mean that dead reckoning can be used where necessary. Indeed, flying a full approach with a timed procedure turn is dead reckoning of sorts. Some approaches have short DR transitions from intersections to the final approach course of an ILS or VOR approach. All you need is a compass, a stopwatch and some vague idea of wind direction and speed.

Of course, I'm not suggesting that you dead reckon your way through an entire IFR flight plan or that you plot a DR course to an intersection 50 miles away. But it's certainly acceptable to dead reckon to a navaid whose official service volume is just a few miles ahead, especially if you're in radar contact. Keep in mind, however, that if you fly any off-airways route, you're expected to determine a safe enroute altitude on your own.

Notam nonsense

Another regular source of the unflyable clearance is the host computer which issues ATC's preferred routes. In most cases, it doesn't know (or care) that a facility defining a preferred route is notam'd out of service. The computer will issue and often the controller will read the clearance as though the facility were on the air. In some cases, the flight data controller may revise the clearance to work around the faulty navaid but this is far from standard practice. Right from the start, you'll be issued a clearance that can't be flown electronically in case of comm loss. The expectation, and the general practice, is that the clearance will be amended enroute, either via radar vectors or to suitable navaids.

What to do? If you're uncomfortable in accepting a clearance with a known gap , discuss it with clearance delivery before you depart. There may be a delay, but the delivery controller will work out a route that satisfies you.

Otherwise, go ahead and depart, planning to mention the discontinuity to ATC further down the line. It's unlikely that you'll lose comm before that happens, but if you do, just fly the route as best you can, even if that involves a DR leg. You can bet ATC will move any traffic out of your way.

• Section Three •

Getting the Route You Want

Expediting Your Flight

In the first section we got an inside look at how air traffic control operates; now it's time to get to the nub of the matter: how to make the whole complex institution that is ATC work for you.

As we noted in the preface, an instrument pilot can muddle along allowing the system to manipulate him or her and still get to the destination; however, it's likely that a lot of delays and hassles will be involved.

Alternatively, the pilot can do everything possible to take charge of the system, pushing the right buttons to get it to do what the pilot wants. It's not always possible, of course, but there's a lot the pilot can do to improve his or her situation.

First, IFR *Editor-in-chief and charter pilot Paul Bertorelli offers some good advice on how to make use of that increasingly common cockpit gadget, the loran (or GPS) receiver—even if it doesn't sport the coveted IFR certification needed to make it a "legal" navigational instrument. While using it as the sole means of instrument navigation is a definite "don't," there's a surprising amount that you* can *do with that "VFR Only" nav receiver.*

Flying IFR With a VFR Loran

"Nice loran. Got the beacons plugged into the user database yet?" I asked, settling myself into the right seat.

The pilot looked genuinely puzzled. He'd just bought the airplane a month before and hadn't yet mastered all of its various gadgets. He had asked me to give him some dual towards an instrument competency check. We were about to depart into soft IFR for some approach practice.

"Well, actually, I haven't even used it yet," he said.

"No sweat. I'll just punch in the NDBs while you're getting a clearance," I replied.

"Hey, wait a minute, we can't legally use that thing in the clouds, can we?" my student asked.

Now it was my turn to be puzzled. Was this guy serious? Turns out, he was. I later learned that he believed that a "Loran for VFR use only" placard meant you had to turn the thing off upon entering IMC.

At the opposite end of the same spectrum is a very good friend of mine who owns an Northstar M-1 loran, a VFR unit. This guy is an arrow-straight hot stick, a bedrock of the aviation community, yet he routinely files IFR as a /R, implying to an unsuspecting ATC that he's a legal RNAV.

"So sue me," he said once, when I inquired delicately about this minor character flaw.

So, what's a body to do here? Should you be a good scout and resist the temptation to substitute uncertified loran/GPS for honest-to-goodness RNAV? Or is it better to go hell-bent-for-leather and pump that $1500 box for all it's worth? We lean towards the latter, with certain reservations.

What's legal RNAV?

There are lots of ways to file legally as a /R and the chief difference between them is money. Ignoring the INS, Omega and other systems found in heavy iron, general aviation RNAV usually means loran, loran-GPS multi-sensor or rho-theta RNAV.

Rho-theta sets, such the King KNS-80, are sometimes called VOR movers because they use VOR radials and DME distance to electronically displace the station to the desired waypoint. There are a lot of rho-theta receivers out there and almost all of them are approved for IFR en route, terminal and approaches. To find out for sure, check the aircraft flight manual. It should have a supplemental section describing the equipment and there should be a Form 337 filed somewhere in the aircraft paperwork.

Loran has been approved for IFR a lot longer than most people realize. The first IFR-approved sets date back to 1986. Since then, more IFR-legal lorans have come on the market, including King's KLN-88, Arnav's R50i and FMS 5000, Northstar's M-2 and IIMorrow's Apollo FMS.

In order to be approved for IFR, a loran must meet a specific technical standards order or TSO. For IFR lorans of recent manufacture, TSO C60b applies. In addition to the TSO, the local FSDO has to sign off the installation by approving a Form 337. Once all that's done, the loran is good to go for primary IFR navigation for en route and terminal operations and you can legally file as a /R. TSO C60b also provides specs for approach-capable lorans but so far, none are approved for that purpose.

What's not legal?

The vast majority of loran and GPS receivers are VFR-only boxes. Generally speaking (although not always), that means they don't meet any particular TSO, even though the receiver may function just as well as one that does. For Part 91 operations, you can install any piece of equipment you like into your panel, whether TSOd or not, as long as you have a Form 337 describing the installation.

For VFR-only lorans and GPSs, the FAA does publish advisory circulars which describe how these receivers are supposed to be installed. Although not strictly binding, the circulars are considered "semi-regulatory" in some circles and anyone who chooses to ignore what they have to say better have a convincing alternative. For both loran and GPS, the advisory circulars offer specific guidance on installations and require a placard that says "Loran/GPS limited to VFR use only."

Exactly what that placard means is open to debate. In our view, it means that uncertified loran or GPS is not to be used as a sole means of IFR navigation. Period.

In other words, if your VOR heads are in the shop for service and you blast off into the scuzz with only your trusty Flybuddy for guidance, you're asking for trouble. This in spite of the fact that the Flybuddy is nearly as reliable as VOR and, in some cases, more accurate. (We're not talking logic here, just legality.) Should the feds wish to bust you on this rap, they'd probably start with FAR 91.9, exceeding aircraft operating limitations, among others.

As long as the legal, approved equipment is aboard, however, there's nothing to prevent you from relying on loran or GPS as an indispensable aid to navigation. (If you take indispensable to mean nearly the same as primary, that's your prerogative.) So, we see nothing at all illegal about using uncertified loran or GPS to track airways, fly direct to intersections and as a back-up on approaches, especially NDB approaches conducted in IMC.

To file or not to file

I've occasionally heard a pilot wax poetic about his new loran or GPS, only to have some cynic in the assembled crowd observe that "yeah, but it's not IFR approved." The only sensible rejoinder to that is this: So what? Having a certified receiver gives you the dubious privilege of filing as a /R but not much else of real value.

As I pointed out, some pilots file as a romeo anyway. Don't think for a minute that this is news to the FAA. The agency is fully aware that IFR pilots are mucking around in the gray area defined by the phrase "primary navigation."

"We can't get inside a guy's head," one FAA certification official told us. "There's just no way we can tell how a pilot is navigating. We don't operate by osmosis." We doubt if an extensive ramp check would reveal the subterfuge, either, assuming of course, that the flight was uneventful. But if something goes afoul (you bust an altitude, get lost, run out of gas or otherwise screw up), that little /R on the flight strip could prove embarrassing. Or, as one FSDO inspector we talked to said, "Use the loran? Sure, have a nice flight. If things don't work out, we'll be in touch." For what it's worth, we haven't seen any enforcement action on this issue; it's terra incognita.

Nonetheless, it doesn't make much sense to file as a /R unless you've got the approved equipment. The chance of a bust is remote but why risk it? You can get just as much IFR utility out of your VFR box by other means.

Go for it

Even an inexpensive database loran or an older set with capacity for a few dozen waypoints is an extremely powerful navigation tool if used to its utmost. In keeping with a policy to use everything in the panel, I try to set up the loran or GPS before departure and keep it in the navigation loop until touchdown at the destination.

Lately, I've been using a Garmin 100AVD GPS receiver or the junior version, the 55AVD, both in portable mode. For departures, I prefer to set up the GPS for the first fix and rely on it until out of the terminal area. This leaves both VORs free to be set up for the approach back into the departure airport, should that become necessary. With both VORs free, I can set up the entire approach, complete with crossing radials. If crossing radials aren't necessary, you can set up both VORs for the inbound localizer, the missed approach procedure or the first fix specified in your clearance.

One of the most useful features of database lorans is a nearest airport, VOR or intersection mode, accessible with a minimum of button mashing and knob twisting. This could prove to be a lifesaver if you depart into IMC and need to get back into the airport *right now*. In a non-radar environment, you're on your own and being able to punch up the closest intersections or NDBs and proceed direct to the fix upon which the procedure turn is based would certainly be convenient. You can usually accomplish all of this with VORs and an ADF, but it's time consuming. Who needs that when your attention is diverted by an emergency?

Tracking the airway

Despite the widespread use of loran and GPS (legal or not), IFR pilots are usually consigned to trundling down airways defined by VORs. These systems are definitely up to the task but, depending on how they're set up, tracking an airway can be confusing. One issue is lack of accuracy. But the problem isn't with the loran or GPS, it's with the VORs. Loran accuracy is generally one to two-tenths miles, depending on atmospherics and station geometry. With selective availability on, GPS delivers 100-meter accuracy, give or take 100 feet.

By comparison, VOR accuracy varies with distance from the station. Consider this example: Two L-class VORs, with nominal 40-mile service volumes, define a 70-mile airway segment. The changeover point is midway between the two stations. Let's say you're on the airway 30 miles out from VOR A, with a centered needle. If you last signed off your VOR via airborne check, you could have a perfectly legal six-degree tolerance, thus that centered needle might mask a three-mile error either side of airway centerline.

It's unusual for either GPS or loran to become that inaccurate and still function. So, in this scenario, you have centered VOR needles, the controller is squawking that you're off the centerline and the GPS agrees with him. Which would you put your faith in, your "legal" VORs or a crummy VFR GPS, which happens to concur with the controller? Like we said, the VFR limitation on receivers gets pretty ludicrous at times.

Another source of confusion is the difference between bearing and track and how local magnetic variation affects both. Loran and GPS usually display both bearing and track, the former being the magnetic direction to fly to get to the selected fix, the latter being the actual magnetic course or ground track. If you're flying bearing/track on the display, both values should be within a degree or two of each other in order to hold the centerline. When intercepting an airway, say from a vector, fly the intercept heading until you see the bearing that corresponds to the course that matches the airway. Then turn on course and keep bearing and track aligned.

Magnetic variation—or at least the way the FAA adjusts for it—may introduce minor inaccuracies. Here's why: When built, a VOR's 360-degree radial aligns with magnetic north. As variation advances or declines from year to year, the local variation changes but the VORs aren't realigned until the error exceeds three degrees. The variation that the station is actually set to is called slave variation or declination and that's the value that's published in the A/FD. This is why the variation values given for VORs in the A/FD often don't agree with the sectional for the same area.

Loran and GPS receivers typically use a fixed magvar model, based on the most recent survey epoch. The models don't account for slave variation. The upshot is that if slave variation differs by two degrees from the actual local value, that's how much error you'll introduce by substituting actual groundtrack for an airway defined by a specific radial. In the real world, this isn't worth worrying about, given how sloppy the VORs are. However, if you were using loran to track outbound on an emulated airway to a distant fix not defined by another VOR, the error could become noticeable. If you're obsessive about such things, you can probably reduce the error by setting the receiver for a direct route between the two VORs, assuming the box has a trip function. If the receiver had a manual magvar function, you could also punch in slave variation.

Get a heading

Holding to the airways, of course, defeats the efficiency of area navigation. The idea is to go direct, ideally to the destination or, failing that, to intersections along the way. Just how to do this legally with a VFR loran is a point of contention among some pilots. We'll give you several suggestions and you can take your pick.

For sticklers who worry about a FSDO inspector lurking on every ramp, we suggest flying direct off airways routes with a VFR box only when in radar contact. To make things nice and tidy legally, ask the controller for a vector to the fix you're navigating to, then follow the loran or GPS. If the fix is beyond the controller's sector, ask for a vector until you're able to proceed direct. There's nothing wrong with suggesting to the controller a heading that matches the bearing your loran is showing.

A variation on this theme occurs when the controller says "fly heading 230, when able, proceed direct." This is a kind of blank check, routing wise. What it means is that the controller has pointed you in the general direction you need to go; fine tuning is up to you. If you're operating above the local minimum instrument altitude, obstacle protection should be assured but double-check on your chart anyway. If the specified fix is a VOR 200 miles away and you can't receive it, you have three choices: Ask for a heading, fly the loran-suggested bearing or ask for a reroute on the airways. In our view, choice three is pretty silly but you're paying for the gas.

Occasionally, a controller will ask if you have loran aboard. The Boy Scout answer is "Yes, but it's VFR only." This reply indicates that the pilot heard what the controller said but he wasn't listening. Allow us to translate the controller's true meaning: "I'm about to offer you a clearance that will shave 80 miles off the route, do you want it?"

Which brings us back to reality. There are so many loran, GPS and RNAV receivers out there that controllers generally assume that airplanes can navigate direct anywhere. Most don't know or particularly care if these receivers are IFR approved or not, so long as the little flickering datablocks go where they're supposed to. In other words, when you're asked "How are you navigating," don't soil your underpants trying to come up with a clever answer. It's not a trick question.

My favorite method of navigating direct is to simply say something like "Nine Eight Victor requests direct RYMES, heading 330." One of my controller friends suggests this: "Nine Eight Victor requests 330 until we're able direct RYMES." Technically, including the heading turns the clearance into a vector but that's probably an unnecessary bit of CYA. (Just because you're paranoid, doesn't mean nobody's out to get you.)

On the approach

As valuable as they are en route, loran and GPS really shine as supplemental approach aids, especially for NDBs, which need all the help they can get. You have to appreciate this irony: A handheld Flightmate GPS is about 10 times more accurate than an NDB, with its nervous needle hopping around pointing at the spark gaps in the left engine's alternator one minute and a thunderstorm over the horizon the next. But the ADF is legal, the Flightmate is not approved for IFR.

For NDBs, bearing and track navigation (backed up by the receiver's or an external CDI) is the way to go. As with an airway, intercept the inbound course outside the beacon and match bearing and track to the published approach courses. The ADF needle should agree, more or less, but don't be too surprised if it doesn't. A disagreement of more than 10 degrees would be cause for concern, in our view.

In non-radar areas, where you'll have to do the procedure turn, try setting up a flight plan route for the approach, starting the route at the beacon and heading outbound. Most approaches with conventional procedure turns have a 10-mile limit. To establish a point at which to begin the outbound procedure turn, plug in a waypoint on the outbound course four or five miles from the beacon. Upon completion of the turn, resume bearing and track navigation back to the beacon and on to the airport.

If you're being vectored for the ILS, there are a couple of ways to use the loran or GPS to your advantage. I like to set the box to display range and bearing to the outer marker or compass locator. This provides some sense of how the controller's vector is going to work out.

Compare the loran's track (not the bearing) against the inbound course. The angle should be between 20 and 45 degrees. If it's shallower than that as you near the beacon, you may intercept at the marker. (When the bearing and track to the beacon are very close, you're definitely going to catch the localizer at the marker.)

Depending on the wind, an intercept angle greater than 45 degrees means you'd better be ready for an immediate turn inbound because the localizer will come in quickly.

Inside the marker, you can set the box for direct to the airport or to the missed approach holding fix. Again, setting all this into a route function ahead of time simplifies things, since the loran or GPS will sequence automatically.

Avoiding traps

While we think it's smart to get as much of out of a VFR box as you can, there are some traps. The most insidious is the tendency of off-airways area navigation to erode position awareness. At least when you're on an airway, you can glance at the DME and fix position in a couple of seconds. When navigating direct to a fix 300 miles away, it's easy to lose track of exactly where you are.

Last spring, flying from San Jose to Lincoln, Nebraska, with my pal Chuck Kissner, we were given direct Lincoln somewhere over Utah, at midnight. On a handoff a half hour later, the controller said: "Say position." Position? I'd set the Garmin for direct Lincoln. I hadn't a clue. "Malibu One Niner Mike is 756 miles west of Lincoln," I offered. The controller allowed as how he'd like something a little closer so I asked the Garmin for the nearest VORs. Now, for position awareness, I navigate by alternating between the nav page and closest VOR/airports pages and I keep up with VOR/DME cross fixes.

Pilots who are comfortable with navigating direct with VFR boxes sometimes develop lazy filing practices. I'm guilty of this one, too. If you know you can go direct, you tend to file that way, especially if you're phone filing without having charts handy. You just figure you can file direct destination; the computer will probably substitute the preferred route anyway. Unfortunately, preferred routes cover relatively few city pairs so if you file direct where no preferred exists, the computer may stick in a bunch of lat/lon fixes to define your navaidless route. This drives controllers nuts; they have no way of decoding a lat/lon defined route. At the very least, file to a conventional fix inside the departure airport's airspace. Or, better, file a conventional airways route then negotiate something better once you're airborne.

Don't push your luck by flying into a navigational corner and don't let us (or anyone else) talk you into doing something you're uncomfortable with. Always have a plan to back up your navigation with good VOR/DME cross fixes or radar vectors. Remember, the VOR/DME cross fixes are your primary nav. Also, don't forget that loran often quits in the rain and GPS is still under construction; service gaps are common.

The important thing to remember when using this equipment is this: Compared to your junky old VORs, even a non-database loran or GPS is a more capable and utilitarian navigation system. Worries about the legalities aside, these units are real time savers. Don't be afraid to use them.

There's another excellent way to save time, trouble and fuel when trying to get from Point A to Point B: the composite flight plan. While it's true that an instrument ticket increases your flexibility as a pilot enormously, it's also true that flying VFR is always more expedient than going IFR when the weather permits it.

When flying VFR, you can pick your route, as long as it keeps you out of airspace you're not allowed in without a clearance. Theoretically, an IFR clearance can let you do things like going through a TCA as if it weren't there, but in the real world the routing will probably take you around it anyway.

We're not talking about VFR-on-top, here. Instead, a composite is a series of separate flight plans, one of which is VFR—allowing you to take any route you please.

Burbank tracon controller Tom Dray tells how it works:

Two Routes in One

Delays for flow control and "preferential routings" that are anything but seem to be inevitable these days. Often, it seems, the only way of avoiding IFR delays is to go VFR and hope that the weather cooperates. Of course, this carries an element of risk: if you encounter IMC, it may be difficult to quickly negotiate a pop-up a clearance.

If you plan carefully, however, there's no reason why you can't use a composite flight plan to combine VFR and IFR segments. The strategy of departing and arriving IFR with a VFR segment in between can be expanded indefinitely, as the weather allows. And prefiling your IFR portion, for airborne pick-up, will usually reduce or eliminate the delays encountered in getting a pop-up clearance.

Pop-up versus pre-file

Out west, it's fairly routine for an already airborne VFR pilot to request an IFR clearance, especially for an arrival to a close-by airport. In some parts of the country, this is sure to elicit a suggestion to file with flight service but out here, it's usually not a problem.

But there are limits. A few weeks ago, for instance, a twin Cessna called with a typical example of how not to negotiate an airborne clearance:

"Go ahead," I replied.

"Twin Cessna One Two Three Four Lima, off Santa Ana to San Jose, unable to continue VFR, need IFR via direct Gorman, Victor so-and-so to (garble) intersection and the what's-it arrival."

He had obviously done some flight planning, because he rattled off the route in his best big-city, I-can-talk-faster-than-you-can-write tone of voice. Just as obviously, he either couldn't recognize IFR weather until he was practically in it or he just couldn't be bothered to file ahead of time.

A request like this is handled on a workload-permitting basis. For a controller, it involves writing down a bunch of basic information about the flight plan—destination, route, altitude—and passing it on to someone at the flight data position to input into the computer.

If I can do this without degrading service to other pilots, I will. On the other hand, if I'm busy (say the weather is marginal and getting worse and a bunch of pilots are already dodging build-ups and making special requests) I simply may not have the time. It's just not possible to separate a lot of traffic and enter a flight plan at the same time. In this case, I asked the pilot to contact FSS to file.

The twin Cessna pilot obviously preferred to go VFR, probably to save time. But if he had gotten a good weather briefing and had even *suspected* he would need an IFR clearance, he could have filed a flight plan to cover that eventuality, planning to pick it up well ahead of encountering IMC. If the weather cooperated, he'd have the option of going VFR or IFR. If the weather went downhill, his flight plan would have been in the system, ready to go. All I'd have had to do is read it, copy the readback and get him on his way, IFR.

Composite defined

As defined in the AIM, a composite flight plan is two or more separate flight plans, specifying IFR for one portion, VFR for another. When I go from Burbank to San Diego, for example, I can file an IFR flight plan to climb to VFR conditions and an IFR arrival into San Diego, to be be picked up airborne. Between these two IFR clearances, I'll fly VFR, weather permitting.

This can be a real time saver because the IFR route is a scenic detour around the TCA. On the other hand, if I can climb to VFR conditions on departure, I can usually get through the TCA VFR with minimum delay. And if the weather is still sour in San Diego, I can pick up my pre-filed flight plan for an IFR arrival.

You can also file a VFR flight plan connecting the two IFR portions, if you want. Between Burbank and San Diego, however, search and rescue hardly seems necessary. If you do want SAR back-up, one option is to file a VFR flight plan from departure all the way to the destination. Activate it with FSS once you've cancelled IFR and been turned loose by ATC, and you're in business.

When filing a composite with FSS, just tell the specialist what you want to do; IFR to VFR, followed by an VFR leg and then IFR again. For the IFR climb to VFR portion, specify a route and altitude that you're reasonably sure will get you above the weather. Just as a suggestion, don't say "IFR-to-VFR-on-top" when you chat with the briefer. "VFR-on-top" is an altitude assignment on an IFR clearance. What you want is an IFR climb to VFR conditions. The two are often confused and in this context, this could result in a major league run around.

Filing with DUAT might be a bit more prone to error because you can't talk with the specialist keying in the flight plans. Make sure to file separate IFR flight plans, one to VFR-on-top conditions from your departure point and another from a feeder fix on the approach you want at your destination. Some FSS specialists may remind you to file a feeder fix but with DUAT, you're on your own.

In either case, for the departure, file a route that will take you to a definite fix along the way, say a VOR or an intersection. Many ATC facilities have a standard routing they use for VFR-to-on-top clearances; if you know it, file it. Either way, have in mind an alternate plan if you're not VFR by the time you reach that fix.

At Burbank, we simply assign a hard altitude above the clouds, and say "report reaching VFR conditions and cancelling IFR." I often just assign the altitude the pilot says he plans to fly after he cancels.

Picking up the clearance

Once you're on top, you can cancel IFR and fly VFR on whatever route you want. You can usually remain with approach or Center for advisories or you can activate a VFR portion of your flight plan if you filed one.

Don't forget, though, you're really VFR here, not VFR-on-top on an IFR clearance. The important difference is that VFR-on-top on an IFR clearance probably won't yield much flexibility in routing.

Nearing IMC, you'll need to pick up your IFR clearance for that leg. If a significant portion of the flight will require IFR clearance, the starting point for the IFR should be a fair piece before you need it. I usually recommend calling for clearance at least 15 minutes ahead of arriving at the point where VFR is no longer workable. This allows you to continue in the right direction if there's a delay in finding your clearance or fitting you into the flow.

Some pilots add something like "PU O/(fix name)" in the remarks section of the plan. This doesn't change the computer processing but it does give the controller some idea of where you'll be calling for your clearance.

In unfamiliar airspace, you often won't know who to call for your clearance. If you're able to puzzle out whose airspace you're in, call the appropriate facility, be it a Center or approach control. Center frequencies are found in the little "rag" boxes on enroute charts; approach freqs are on the plates and on the communications panels of both NOS and Jeppesen charts. If you get the wrong frequency, don't get nervous. The controller you're talking to will give you the right one.

Be explicit about what you want. Not long ago a Comanche called us VFR on his way to Las Vegas. We established radar contact and gave him VFR advisories for 30 miles or so. Just as he reached the limit of our airspace he advised that he was "ready to copy clearance."

"Huh?"

Seems he had an IFR on file. But since the starting point he selected was not in our airspace, we had no way of knowing what he wanted until he spoke up. Once your IFR is activated, by the way, don't forget to close the VFR portion of your flight plan (if you filed one) with FSS. ATC won't do it automatically.

One variation on the composite theme is to file IFR for the entire route then depart VFR, planning to pick up your clearance only if the weather warrants. This will work but if you venture too far from your point of origin before picking up the clearance, the ATC facility you call may not be able to find your flight plan.

New twists

Recent changes to the AIM and the controller's handbook (7110.65 or point 65 in controller-speak) address the subject of picking up IFR clearances when airborne. Unfortunately, they don't seem to agree. For that matter, the way I read it, the AIM doesn't seem to agree with itself.

The AIM says "Any aircraft departing VFR, either intending or needing to obtain an IFR clearance enroute must be able to climb in VFR conditions to the MIA or MVA (minimum instrument and minimum vectoring altitude, respectively) in order to receive the IFR clearance."

That seems pretty cut and dried. But the very next sentence is: "The pilot will also be responsible for terrain/obstacle clearance until reaching the MIA or MVA. Any aircraft that cannot climb, in VFR conditions, to the assigned altitude for IFR clearance and/or accept responsibility for terrain obstacle clearance *should remain VFR.*" (Emphasis added.)

The .65 on the other hand, says that if a controller is aware that a pilot cannot climb in VFR conditions to the minimum IFR altitude, he may issue the IFR clearance if the pilot agrees to accept responsibility for terrain and obstruction clearance until reaching the MIA. It emphasizes that the controller must avoid issuing specific course guidance below the MIA if it could even imply that ATC is providing terrain separation.

In my opinion, if you are at 3500 feet in an area where the MVA is 5000 feet, I can legally issue a clearance such as: "Cleared to San Diego via Van Nuys, V186 (etc.), maintain 5000." How you get to VNY and make the climb to 5000 without running into terrain is up to you. Making sure that there won't be any other IFR airplanes around is up to me.

I have to point out that there's a fair amount of disagreement in the controller ranks about this. Some feel that it's illegal, or a shirking of their responsibility to issue a clearance that saddles the pilot with responsibility for terrain avoidance. Personally, I'm a little uncomfortable with the idea, but I accept it as legal.

I would suggest, however, that asking for or accepting a clearance below the MIA or MVA is something to be approached with caution, and avoided altogether unless you are very confident of your ability to stay out of the weeds.

When things don't work out

On rare occasions, the pop-up and perhaps even the composite will lead to just the kind of delay you hoped to avoid. It happens. It may be better to land and take your delay on the ground. All things being equal, controllers work on a first-come, first-served basis applied in the spirit that somebody who pops up usually gets in line behind the pilots who may already be waiting on the ground. Still, the farther you can go VFR, the less exposure you have to the vagaries of ATC.

The bottom line is this: If you're going to play around with airborne pick-ups and pop-ups, have a comfortable fuel reserve, especially if the weather is crummy and the system is choked with traffic.

Also, there are some situations where this system would not be a help at all. For instance, IFR from San Diego back to Burbank is a straight shot through two TCAs, so IFR is often faster than VFR. Inbound to LAX, flow control may apply to *any* arrival, and a request for an airfile will be turned down flat.

Some controllers, I might add, see airborne pick-ups as a kind of "butting in line" and become resentful. Personally, I'd much rather work you VFR as long as possible rather than hold your hand IFR all the way when it isn't necessary.

How sweet it is

The last time I went into San Diego on a composite, the weather there was zero-zero. IFR flights into San Diego were being held on the ground at their departure points. So I launched VFR, with lots of gas and a place to divert if the fog didn't lift. Over Oceanside VOR, about 30 miles north of San Diego, I called approach.

"San Diego approach, Cessna Three Five Eight Eight Lima, over OCN at 5500 feet, prefiled for IFR pickup to Lindbergh, we have Charlie."

San Diego came back with a squawk, ident request and radar contact, followed by:

"Cessna Eight Eight Lima, Lindbergh RVR just coming up to minimums. Cleared to Lindbergh via present heading, vector to the localizer. Keep your speed up, you're number one."

It's not always so sweet but as long as you understand the pitfalls, you'll get a break a lot more often than you might imagine.

In the last section, Tom Dray described how not to ask for a clearance while enroute. Nevertheless, airfiling and pop-up clearances are extremely useful tools for the IFR pilot and can save a lot of trouble. For example, sometimes filing IFR just doesn't make sense, so VFR with flight following is the order of the day. However, it's often useful, and sometimes necessary, to get hooked back into the system towards the end of the flight. Knowing just how to negotiate a pop-up clearance is a valuable skill, one not often practiced by the average instrument pilot.

IFR *edtior Paul Bertorelli, along with controllers Paul Berge and Denny Cunningham now cover the niceties of obtaining a clearance on the fly.*

Asking For a Pop-Up

ATC clanks along on mid-1960s computers, suffers and survives the weather just as pilots do and at the end of a good day, with any luck, the system has herded its airplanes to the right airports while remaining at least one flight strip away from total collapse.

Not that I'm complaining. It's a great system, compared to what Europe and rest of the world has to put up with. Still, when time is of the essence (when isn't it?), it sometimes makes more sense to fly a trip VFR, connecting with ATC only to the extent necessary to allow a full-up, IFR flight plan if things get ugly, weatherwise or otherwise.

Around some megaterminals I've flown to, this isn't so much a strategy as it is a survival skill. Unless you like sitting on the ground or flying 60 miles out of your way, VFR with advisories is the way to go, with an airfile or pop-up as a back-up plan.

Terms defined

Some pilots use the term airfile and pop-up interchangeably. While they do accomplish the same thing (you hope), they're distinctly different. I like to think of an airfile as the utmost in self-flagellation. It's simply a standard flightplan filed with FSS over the radio, while airborne. Since you have to provide data for all 16 blocks on the flightplan form, airfiling is even more tedious than phone filing. Sometimes it's the only way, though.

A pop-up, on the other hand, is an on-the-spot request for a clearance directly with ATC, using what the Pilot/Controller glossary defines as an abbreviated flight plan. Usually, a pop-up is an approach clearance but depending on who you're asking, it can be more. Much more.

There are several schools of thought on when and how to use pop-ups. Some say resorting to a pop-up is a symptom of shoddy planning while others argue that it's okay to consider pop-ups as the standard way of doing business. Take your pick. Although I tend to bridle at the mere suggestion of regulation, no matter how reasonable, I favor sparing use of pop-ups, if only for self-preservation.

On the other hand, I use airfiles (pre-files, really) regularly when the weather is decent VFR and going IFR is just too big a pain in the butt. Departing VFR and picking up a clearance en route is sometimes the only way to beat the preferred route system. If you can talk your way into radar advisories before getting your clearance, so much the better; you'll get virtually the same service without getting sent around the horn on a preferred route.

A good example is departing Washington National, eastbound for New York. DCA is a busy airport and it sometimes gets behind on IFR releases. But the tower pumps out the VFRs without delay, with a turn down the Potomac and a direct heading out of the TCA. From there, you can proceed VFR or pick up an airborne clearance just beyond the TCA.

Even if you go VFR for the entire trip, there's enough radar around to assure uninterrupted traffic advisories without the nuisance of roundabout routing. Yes, I've heard the argument that VFR advisories aren't as safe as an IFR clearance but I've also noticed that ATC doesn't exactly promote radar assisted collisions between the VFRs. I'll take my chances on VFR-with-advisories and pocket the gas money, thanks.

When the weather sours

Let's consider that Washington trip again, with marginal VFR or IFR weather at the destination. In this case, I still think it's worth the few minutes saved to depart VFR. But before departing, I'll pre-file an IFR flight plan for an airborne pickup somewhere along the way. Weather permitting, I try to specify an airborne pickup at a fix and altitude that's least likely to give ATC fits. (Such that it's possible to know such things.)

If you haven't had the foresight to pre-file, an airfile may be the only choice, albeit an inefficient one. It's awkward to file over the radio and there's always the chance the flight plan will go to the wrong facility entirely or get delayed in transit while you're trying to dodge clouds.

Filing while airborne is not much different than filing by phone. On a busy day, the biggest problem may be finding a quiet FSS frequency. Ask for a discrete freq; don't use 122.2 to file. Normally, FSS prefers a proposed departure time that's at least 30 minutes after the filing time. For airfiles, however, just specify a p-time that's a few minutes later than your filing time. The FSS computer should forward the flight plan to the appropriate ATC facility almost instantaneously.

Before signing off with FSS, request an ATC frequency for the area where you plan to pick up the clearance. It's quite likely to be different from the freq which appears on the en route chart and if you have to play who's-got-the-flight-plan, the more frequencies you have, the better.

Pop-ups

Negotiating a pop-up is, to understate the matter, an uncertain art. But you don't have to be a towering intellect to know when it'll work and when not. If the frequency is wall-to-wall and the controller sounds like she's one synapse away from a trip down the tubes, don't bother.

If you can remain VFR comfortably, try an airfile or land, have lunch and do it over the phone. On the other hand, on a quiet Center frequency, it's best to keep your pen cocked; a pop-up request may yield a clearance across several states.

That's exactly what happened to me and my friend Todd Huvard on a flight from Texas to Las Vegas a while back. Todd is your classic, cigar-chomping, self-made capitalist business pilot but he's not known for his subtle understanding of ATC procedures. We were VFR just northeast of Phoenix at 10,500 feet. As dusk descended, Todd idly asked Albuquerque Center for a clearance to LAS.

Being schooled in the ways of ATC, I expected (a) "make that request with the next controller" or (b) "contact Flight Service..." or (c) gales of laughter. But without a moment's hesitation, the controller came back with a route, altitude and a squawk code. Instant IFR. Huvard says he's accustomed to that kind of service.

That leg was about 300 miles, or about the same distance Boston is from Washington. Now, I'm not too shy to ask for a pop-up from Boston to Washington but I believe it's considered a felony in any state east of Ohio.

Airborne clearance

Whether airfile or pop-up, you have to meet certain conditions in order to obtain a clearance. Radar contact is not necessarily among them, although it certainly helps. If you're already in the system on advisories and squawking a discrete code, IFR clearance delivery is greatly simplified.

If you're calling ATC cold, say for a pop-up approach clearance or a short en route segment, you can still get in the door. It may take a few minutes longer because the controller will have to figure out where you are and assure that separation can be maintained before your clearance is activated. If you're below the minimum en route altitude, minimum instrument altitude or minimum vectoring altitude (MEA/MIA/MVA), you can still get a clearance but you'll have to provide your own obstacle and terrain clearance. This last point reflects a subtle but important change in airborne clearance procedures. Until late 1992, the AIM said quite plainly (para 4-88) that "any aircraft that cannot climb in VFR conditions to the MIA [to receive a clearance]...should advise the controller that an emergency is being declared."

The controller's manual, on the other hand, has always had language that allows a controller to issue a clearance below the MIA, as long as the pilot is providing his or her own obstacle clearance. The FAA has tried to close this loophole by amending the controller's manual to prohibit clearances below the MIA.

Thankfully, that proposal was quietly dropped later in the year. In October of 1992, the FAA brought the AIM into compliance with the controller's manual by amending paragraph 4-88 to allow pilots to accept clearances below the MEA/MIA/MVA. The phrase about declaring an emergency has been deleted.

Should you decide to accept a clearance while operating below the

MEA/MIA/MVA, you're expected to meet the minimum altitude requirements described in FAR 91.177. Technically, the minimum altitude requirements apply to airways and published route segments, which have published minimums. When you're operating off airways, with no published minimums, you're supposed to allow 1000 feet vertically and four miles horizontally from the highest obstacles. In mountainous areas, the vertical requirement is 2000 feet.

Is it legal?

There's one other legality about pop-ups we should mention, although I'm not sure it's worth worrying much about. FAR 91.169 gives a laundry list of information a pilot is required to provide when filing an IFR flight plan; these items are automatically included when you file with a standard FAA flight plan form.

When you file a pop-up, the controller accepting your flight plan is interested mainly in your tail number, type and equipment code, destination, altitude and route. Some will ask for souls on board and fuel. He or she could care less about the pilot's name, aircraft home base, time en route and other details, including an alternate, if one is required.

Does this mean that you are therefore relieved of the responsibility of providing ATC with this information? Maybe. Maybe not. The first sentence of FAR 91.169 says: "Unless otherwise authorized by ATC, each person filing an IFR flight plan shall include in it the following information." The laundry list follows. We take that to mean the rest of the information isn't required, as "authorized by ATC."

As usual, there's a catch. In a 1981 case, the FAA violated a pilot for (among other things) failing to declare an alternate when he filed an abbreviated flight plan with ATC. In his defense, the pilot argued that, as defined in the Pilot/Controller glossary, an abbreviated flight plan "includes only a small portion of the usual IFR flight plan information." It goes on to say that "it is frequently used by aircraft which are airborne and desire an...approach...or aircraft on the ground [which] desire a climb to VFR on top."

In the 1981 case (Administrator v. Gately), the law judge went his own way in interpreting the regulation. In finding against the pilot, the judge ruled that not including the alternate in an abbreviated flight plan is "no way to conduct IFR operations. It is a violation of the regulations. I don't care how many people get away with it."

Even though the unfortunate precedent exists, we still think pop-ups are perfectly legal, given the definition given in the Pilot/Controller Glossary and the escape clause in FAR 91.169. Since the Gately case, we've seen no pattern of FAA enforcement on abbreviated flight plans.

Still, if you're nervous about getting nailed, just give the controller an alternate. He won't even bother to write it down but at least it'll be on the tape if your pop-up doesn't work out as planned.

Pop-ups from ATC's view: no problem...

Lots of fine words are said about pop-up flight plans, but really, how much trouble are they for ATC? There are three possibilities for your flight plan status: ATC has never heard of you, you're already getting VFR advisory service, or you've pre-filed a full flight plan and just want to pick it up in the air.

Obviously, "never heard of you" will require the most ATC effort to get you into the system, but it's still not all that complicated. As with most other situations, all ATC wants to know is 1) Who are you?, 2) Where are you? and 3) What do you want? Let's suppose you're VFR in Cherokee 1234A, 50 miles southeast of the Panoche VOR at 6500 feet and requesting an IFR clearance to Oakland.

In response to your initial ID-only call, the controller will probably punch the CODE key on his console, enter your callsign and have the computer assign you a squawk. Chances are, you'll want radar service of some kind so he might as well start out ahead.

After obtaining your request and assigning the squawk, he has to make sure you're in his airspace before issuing the clearance, or coordinate with another controller as required. He'll also need to check your altitude to see if you're at or above the MIA for the area. If not, he's required to ask if you can maintain your own terrain clearance during the climb to the MIA.

Once the radar system sees your squawk, the computer will automatically tag the target with your ID and Mode-C altitude, if available. You'll then be "Cleared to Oakland airport via direct Panoche, V301, maintain 8000 feet." To finish up, the flight plan must be put in the computer by pressing the FP key and entering: N1234A PA28/A 130 PXN133050 E2201 080 PXN133050.. PXN.V301.OAK.

That's worst case. There are actually shorter ways to do it. In any event, the whole thing can take less than a minute for the typical short-range clearance. In a pinch, a supervisor or assistant controller can make the entry, if necessary.

Most likely, an aircraft receiving VFR advisories will already have a flight plan in the computer. All that's required is the assignment of an IFR altitude and maybe a route amendment to fit into the IFR structure. If you're expecting to need a clearance somewhere down the line, asking for advisories is a good way to get your toe in the door.

The process for using a pre-filed plan is halfway between "never heard of you" and "already on advisories." The flight plan, which will already contain a squawk, must be activated, a radar track started, the aircraft's current position updated if not the same as the pickup point shown in the flight plan and the route checked for acceptability. Probably less typing than for a "never heard of you" aircraft but, especially for short flight plans, not really a lot less work.

The point of all this is if you make a habit of pop-ups, use the strategy that best suits your situation. You'll save yourself and the controller a lot of grief.

...but ask nicely

A pop-up is much like having guests pop-in at dinner time; success depends on the relationship between host and guest. If the guest is obsequious enough, "...I could leave, really!" and the host is gracious, "...no trouble, really!" then the evening is a big hit. If, however, Aunt Martha and Uncle John continually drop in at the same time with, "Jeez, we're starved; where's mine?" then one day the host will crack.

True, in the ATC environment the host/controller is being paid by the guest/pilot. But certain rules of pop-up etiquette should still be maintained. A pop-up clearance should never be viewed as a way to "beat the system."

Think of it as a means to fully utilize the system's potential. In approach airspace, a pop-up requires no more controller time and effort than does a pop-up VFR.

In both cases, a pilot needs to enter approach's airspace—either ARSA or TCA—and land. In both cases, the controller is not expecting the call, but is prepared. The pilot is radar identified (usually) and given a clearance and vectors. The VFR arrival is given the same service less the clearance limit and IFR separation.

Workload-wise, it's a draw. The hardest pop-up to work is the pilot who waits too long and desperately demands an IFR clearance after getting into the crud. An approach controller up to his eyeballs in IFR traffic may have no place to put an IFR pop-up regardless of how deep the pilot has penetrated. In approach airspace, if you think you're going to need a pop-up, better to ask sooner than later. And you can't reasonably expect more than an approach clearance from a tracon controller. If the controller considers your request too complicated (i.e. requesting Dayton to New York) you may be asked to contact FSS and file.

Approach may ask for an abbreviated flight plan. All the controller wants is the basic poop for the ARTS computer—namely: callsign, type aircraft and equipment suffix (the suffix they don't always care about), destination, altitude, and route. Please skip the souls on board, color of aircraft, TAS, phone numbers, next of kin, and credit card expiration date. Approach control will handle you for a very short while, assuming you land within their airspace.

If you wish to pop-up and file to a destination outside approach's airspace, then the controller may tell you to call Flight Service and file with them. The approach controller cannot process the required flight plan information of souls on board, alternate, aircraft color and so forth. Again, it probably won't matter if the flight goes smoothly. But that's the way with all short cuts; if no one complains, who cares?

Pop-ups are a good way to enter the system and get an approach. They are not, however, a good way to avoid filing a flight plan with Flight Service.

Though instrument pilots are trained to fly in less-than-perfect visibility, in the real world VFR conditions are the rule rather than the exception.

Taking advantage of your ability to see your destination can save a surprising amount of time for both the pilot and controller. Controller Paul Berge explains how.

Visual and Contact Approaches

Not every IFR flight need end with an instrument approach. If the weather's fine, you can always cancel, squawk 1200 and arrive VFR: this is a particularly good method when arriving at an uncontrolled field, since it lets you get out of the system early and devote all your attention to fitting in with the flow of traffic at the airport.

Alternatively, the controller can issue a visual approach, or you can ask for a contact approach. Both these methods can help you get to where you're going a lot faster than flying a full-blown approach.

Visual approaches

The visual approach is the easiest approach for the controller to issue and when the weather is good, it speeds things up.

It's not an instrument approach and therefore the visual approach saves us the need to spell out the altitudes to maintain until established. Also, there's no missed approach. The aircraft goes in and is not expected to return.

Should the pilot lose sight of the airport or, for whatever reason, not make it in, he now becomes a simple go-around with alternate instructions created on the spot, "..radar contact, cleared to so-and-so, fly heading 360." The go-around is treated like any other pop-up clearance.

But once cleared for the visual, you're still on an IFR flight plan until you cancel.

In order to vector an aircraft for the visual approach, the field must be reporting weather that's better than basic VFR. Visibility must be 3 miles or better and the ceiling at least 500 feet above the MVA. If the visibility is 3 miles, but the ceiling less than 500 feet above the MVA, the controller then vectors the arrival for the instrument approach. Should the pilot subsequently see the airport, the controller may then clear the arrival for a visual approach.

Often, when the ceiling is too low to vector for the visual approach, yet still VFR, we may tell a pilot "fly heading 230, vector ILS 12L final approach course; if you see the runway, advise." In that case, we can take the pilot close enough to take a look and still have him heading for the localizer.

At satellite airports, where weather reporting is not available, the controller's judgment is relied upon to decide if a visual approach might work. The controller is allowed to use pilot reports and weather from nearby stations in making this call. It's up to the pilot to advise of conditions and call the airport in sight or reject the visual and request an IAP.

This is one instance in which a Center controller is allowed to use visual separation. The controller may clear an aircraft for a visual approach, provided he or she has met the requirements of "resolving potential conflicts; the aircraft is number one in sequence; pilot reports the airport or the preceding aircraft in sight and can follow or...."

You could land at an uncontrolled airport, having not yet cancelled your IFR flight plan and have another IFR plane land behind you, also on a visual approach. On any visual approach clearance, the pilot assumes *all* responsibility for terrain clearance. We therefore assume that you can see the 2000-foot antennas the city council allowed MegaBucks Communications to build off the end of the runway, and you will avoid them. Don't expect ATC to say, "Cleared visual approach. Do you see everything?"

It's also the pilot's responsibility to conform with the local traffic pattern. Once a pilot reports the preceding aircraft in sight, and is told: "Follow the heavy DC-8, cleared visual approach," it becomes the pilot's responsibility to provide spacing and wake turbulence separation. Of course, in the case of following a heavy, we'll throw in the watery caveat, "caution wake turbulence." This gets us off the hook in court.

If you see the traffic called by approach but are in no position to follow, inform the controller nicely: "Hey, swell job, approach. I see the guy but can't possibly follow." We'll prove understanding and do our best to find someone even more difficult for you to follow.

The contact approach

A distant cousin to the visual approach, the contact approach is a well-guarded secret. More the black sheep in the family, surprisingly few pilots are familiar with the procedure and perhaps that's as it should be. Like the visual approach, the contact approach is not an instrument approach and it has no missed approach. Like the visual approach, it requires the pilot to see outside, although not much outside.

The AIM says this:"...an aircraft on an IFR flight plan, having an ATC authorization, operating clear of clouds with 1 mile flight visibility and a reasonable expectation of continuing to the destination airport in those conditions, may deviate from the instrument approach procedure and proceed to the destination airport by visual reference to the surface." In short, this is an ATC authorization to scud-run on an IFR clearance.

All the pilot must see is the ground or at least enough of *something* to fly towards the runway. When flying my 65 horsepower Champ, I feel comfortable doing this; in a C414, the same maneuver might turn ugly, This, of course is the pilot's call.

To make the contact approach less inviting, the .65 includes these restrictions. First, the pilot must request a contact approach. "It is not," says the book, "in any way intended that controllers will initiate *or suggest* (my emphasis) a contact approach to a pilot." That sounds fairly clear, but sometimes, pilots get a little confused about what they can and can't do. On one occasion, for instance, when the ceiling was 900 broken and the visibility 10 miles, one pilot on my frequency kept requesting a visual approach. I'd dutifully read him the weather followed by, "...the field's below VFR, say alternate request."

"I can see the whole airport! What I want is a visual approach," said the frustrated pilot, for the third time.

"Roger, I can't issue the visual. *Contact* me on 118.6 with alternate request." The light came on and the pilot requested the contact approach. I don't think I conformed with the spirit of the regulation but no one sued.

Ground visibility required

To further restrict the contact approach, the airport must have "reported ground visibility." This means official visibility from tower or NWS or other licensed observer, not Uncle Fred who runs the Unicom. One other requirement: the airport has to have an instrument approach;

can't do a contact approach to a dry river bed, at least with engines running.

While the contact approach has no missed approach segment, the controller must issue an alternate clearance if there's a chance you won't make it in. My advice: Unless you're familiar with the procedure and the terrain, stick with the published approach.

When you're considering when and how to use any kind of visual separation (including VFR climbs and descents and VFR-on-top), remember, instrument procedures exist to maneuver pilots where they can see the runway. Much beauty exists outside the aircraft in flight. Even though ATC has you in a vice lock, stop and smell the roses; look outside; enjoy the view, and if you see the airport and feel like taking over visually, say so. It becomes just like real flying.

Handling Preferred Routes

No matter how flexible you or your equipment may be, you still have to deal with ATC's route structure. You may always want to go direct, but chances are the computer will give you the same route no matter what you ask for.

Welcome to the preferred route, which, though it sounds like it's there to serve you ("Would you prefer a window seat, or an aisle?") is really there to make life easier for ATC ("We'd prefer it if you were to take V134...in fact, we prefer it so much we'll give it to you no matter what.").

Often, there's nothing you can do to avoid a preferred route when filing, even if you are aware of it (not all are published). There are, however, many tricks of the trade that can be used to beat the system. Unfortunately, there are also many ways for these workarounds to backfire.

Editor Paul Bertorelli illustrates how the system works and some of ways to make it work for you.

Beating the Preferred Route

Late last summer, I had occasion to fly two round trips on the same day from our homebase at Westchester County to Williamsport, Pa. On the L-25, it couldn't be simpler: Direct Sparta, then V188. It's one of those deceptively logical routes that slips right into ATC's computer without arousing suspicion.

Of course, it never works. Sensible as it seems and unbeknownst to anyone who hasn't flown it before, that route runs square into a thicket of kerosene burners funnelling into the approach gates for Newark. It's the last place a controller wants to see a 150-knot Mooney.

Lower and slower westbounds go on the tourist route; a dog-leg that jinks almost due south before turning northwest thence to Williamsport and points beyond. It's no big deal, really. But with avgas at $2.25 and climbing, I'd just as soon not fly the 30 miles the dog-leg adds.

My complaint about the circuitous routing elicited a sincere but puzzled response from a controller friend who works in the New York tracon.

"Why didn't you file for 9000?" he asked. "You coulda got direct Williamsport."

"So how I am supposed to know that?," I replied, explaining that I was never *offered* 9000.

"Well," he said, "I know you can find that stuff, I've seen it published somewhere."

He's right. It is published. We've all seen the preferred IFR routes printed in the back of the A/FDs and furnished by Jeppesen as part of a standard subscription. What the A/FD doesn't explain, however, is how all those canned routes really work and what to do when they take you where you'd rather not go.

What it all means

Preferred routes are used in varying degrees throughout the U.S. but they're most intensively employed in the northeast, where some 1,200 routes exist. Not all of these are published but, nonetheless, the northeast A/FD devotes a full 42 pages to preferred routes, compared to a mere 6 printed in both the southwest (which includes the Los Angeles Basin) and the east central region, which covers Chicago. The northwest region has exactly two pages of published routes.

Although it doesn't always work out that way, preferred routes are supposed to expedite traffic by organizing the flow into fixed patterns. Routes are frequently segregated by aircraft type and/or speed, thus twins flying from Atlantic City to New England can expect the Shark Route, V139 over the Atlantic while singles are more likely to be assigned an inland route.

Preferred routes are derived via the FAA's version of the smoke-filled room. ARTCC A dickers with tracon B over what routes airplanes will fly between cities in each jurisdiction. When everyone's happy, a deal is struck; its terms become a Letter of Agreement between A and B.

Some letters are broad, granting controllers flexibility to move traffic more or less as they see fit and to ad lib routes to suit conditions. Other letters—especially those between two busy areas—may be quite restrictive, requiring controllers to hand off traffic to a neighboring facility only over a certain fix or at a specified altitude. Letters of

Agreement come and go, often without pilots ever being aware of them. If the FAA wants pilots to know about a route, it will publish it in the A/FD or otherwise make known the existence of the route.

The A/FD describes two kinds of canned routes; preferred routes and tower enroute control routes. Although there's technically a difference, the two have come to mean the same thing, at least in the northeast. A complete preferred route—called a preferential departure/arrival route or PDAR is a city-paired clearance that takes an aircraft from one airport to another via a departure fix and a destination fix. For example, a Navajo flying from Buffalo to Washington probably gets V33 BFD V170 V93 BAL, the agreed-upon PDAR.

How the airplane gets to the departure fix after takeoff and from the arrival fix for the approach is up to the Buffalo and Washington tracons but all parties in between (except maybe the pilot) have agreed to accept this routing. There's no guarantee that the route won't get trashed 15 minutes into the trip, but at least the initial clearance gives the illusion that the agreements will be honored.

Tower enroute control routes are kind of a sub-species of preferred routes. They're also city-paired, sometimes using the same airways specified in low-altitude preferred routes. It's a common misconception that TECs were invented during the 1981 controller's strike as a means of routing an aircraft from one approach control to another at a low altitude, avoiding hand-offs to ARTCCs. TECs were widely used for that purpose during the strike but they pre-date it. Following the strike, tracon boundaries were expanded so much that TECs have become routine and, in some cases, rather long. It's possible, for instance, to fly the 400 miles from Boston to Richmond on a TEC at altitudes up to 10,000 feet. And don't be too surprised if you talk to an ARTCC along the way.

TECs have become so widespread that's it's not unusual for one to pass through Center airspace, despite the FAA's intent to hand off airplanes from tracon to tracon.

What you see is what you get

When you file a flight plan, only a portion of what you read to the FSS briefer makes it to ATC. Here's the list: tail number, aircraft type and equipment, proposed time off and departure point, destination, route and altitude and remarks. Centers also get filed TAS. The rest of the data in those 16 blocks, including your filed alternate, remain in the FSS computer.

Almost all clearances originate in a computer that, sad to say, pays little attention to the route you request. Rather it looks at your departure and your destination, and, if a city-paired preferred route or TEC applies, that's what will be assigned. A flight progress strip describing the route is then printed for the controller. Should you happen to be flying where no preferred route applies, the computer will likely route you via a preferential departure route (PDR in ATC-speak). Think of a PDR as a truncated route that takes you just outside the departure tracon's airspace.

At the arrival end, it may toss in a PAR (preferred arrival route) if you're going into a tracon that insists that all aircraft arrive via an agreed-upon route. The segment in between is up for grabs. It may be what you filed or it may be a segment of a preferred route.

All this is not to say that no humans look at the route you asked for. The routes are checked manually but unless there's a compelling reason to do so (weather, congestion, etc.) the flight data specialist or controller responsible for the review probably won't suppress the computer-generated route, at least initially.

A real-life example

To illustrate how this works, I filed two flight plans from White Plains to Martha's Vineyard. The preferred route is CMK V3 HFD TMU V374 MVY. It's not a bad route. However, in a single, I don't like the overwater segment on V374, especially at night or during the winter, so I requested an inland route and specified "NO V374" in remarks.

As it turned out, I got the canned route anyway. If no preferred route exists, the computer might review and even approve what the pilot requested. But it can't interpret the remarks explaining why you don't want the preferred route.

After the strip was printed, a flight data specialist reviewed the route and read my remarks. She didn't change the clearance to honor my request but she did pick up the fact that I'd filed the wrong altitude for that route so she penned in the correct altitude: 7000 feet. Had I called for clearance, I'd have been read this exact routing. By the way, the remark "no V374" isn't much of an attention getter. I should have said "no overwater."

Getting another route

Although pilots would prefer to think ATC works the same everywhere, it doesn't. In the New York area, you can expect to be assigned the preferred route almost always, no matter what you file. Elsewhere, fewer preferred routes may exist or, by Letter of Agreement, controllers may have more flexibility in working out pilot-requested routes.

In the northeast, you might (with emphasis on might) stand a better chance of getting a non-preferred route approved if you at least research the A/FD enough to include some preferred route segments in your flight plan.

For example, flying eastbound out of Baltimore towards Boston, the preferred single-engine route goes up the Delmarva peninsula. If, for whatever reason, you wanted another route, at least file the preferential departure route then branch off at a fix further along the route. At the arrival end, try to file the preferential arrival route.

How can you recognize these routes? Preferential departure routes usually end just outside a radar facility's airspace, often at the departure fixes shown on SIDs. For the arrival fixes, check the STARs.

Maybe such a cobbled up route will be approved, maybe not. Experiment a little. You'll at least learn something about how your local airspace is organized.

One other thing to try is an altitude higher or lower than your ideal, if winds or freezing levels aren't a factor. On my trip to Williamsport, for instance, the A/FD gives the TEC route ceiling as 8000 feet. New York Center owns the airspace above and one of its departure fixes is COATE intersection, just northwest of Sparta VOR. COATE appears prominently in the preferred routes in that direction and is also on the Newark Three vector SID. Had I filed for 9000 and direct COATE, I could have gotten a more direct airways route to Williamsport.

Sit up and beg

If you take A/FD research too seriously, filing a flight plan could get to be like a Ph.D dissertation. Sometimes it's just easier to file the preferred, if for no other reason than the minor satisfaction of copying "cleared as filed." You can always try to negotiate something better once airborne. That's what most of us choose to do.

Wheedling a better route out of ATC is an uncertain art. Sometimes controllers hand out amended direct routing without being asked. On a bad day, they don't even want to talk about it. Where, when, who and how you ask has some bearing on the outcome. Under some circumstances, it's advisable to ask the clearance delivery controller to work out an amended route before you accept IFR release. Let's say your clearance calls for 8000 feet but you want to stay at 6000 to avoid icing. If you're not going to get the lower altitude, better to find out about it on the ground than to try and negotiate lower with an inch of building rime.

In general though, asking for reroutes through clearance delivery is time consuming, both for pilots and controllers. The request has to be bumped up the line (usually by telephone) to the controller actually working the traffic, then bumped back down through clearance delivery while the pilot idles on the ramp.

If you can't accept any but your filed route, you can try to contact the clearance delivery position by phone to find out if there's any chance of approval and how long the delay will be. Actually, you could even call clearance before filing to find out what routes are likely to be approved.

When making airborne reroute requests, remember that controllers don't always have carte blanche to ignore preferred routes. A controller almost always has to coordinate your request with another controller, either in the same radar facility but often by phoning another tracon or Center. This takes time so the sooner you make your request when handed off to a new sector, the more likely you are to get it approved.

Keep your requests as palatable as possible. On a busy frequency, I once heard a pilot ask for a new route defined by several airways and a half-dozen fixes spanning two states. Subjected to this spiel, I wasn't surprised to hear the controller respond with a frequency change and a polite "make that request with the next controller."

Ask for new routes in little bites instead of giant gulps. Be flexible on altitude and/or ask if higher or lower will yield a more favorable route. Some knowledge of the local airspace and traffic flow gathered from the preferred routes will help you tailor realistic requests. Better yet, visit your local radar facility. An hour plugged in on-position is instructive and the controller is likely to pass on a trick or two.

The pop-up solution

In very busy (or just plain weird) airspace, reroutes sometimes won't work no matter what you do. In these circumstances, your options narrow. When the weather is IMC, you just have to soldier along and go where you're told. In VMC, a request for VFR-on-top or a VFR climb to another altitude might produce results but in the northeast, it's unusual. Canceling or flying the trip VFR in the first place is always an option. It's the rare day when VFR isn't quicker than IFR.

When flying in good VMC above an undercast, canceling IFR and continuing VFR with the plan to pick up a pop-up approach clearance at your destination is a possibility. It's a strategy that ought to be used sparingly, if at all. For one thing, you need to be sure that you can remain VFR and that your destination weather is above minimums. Otherwise, you could find yourself stuck on top with no place to go. Pop-ups have variable probability of success, depending on where you are and when. On some parts of the west coast, they're encouraged but in other areas, it's chancy. One trick is to air file a flight plan and pick up a clearance just outside the terminal airspace you're headed into.

It's reasonable to expect a pop-up when flying from a busy area into an uncongested terminal but don't try it the other way around. Blundering into an airport like Teterboro at 4 p.m. on a Friday afternoon and asking for an approach clearance is a lame strategy. There may be so much traffic that a controller will have no way of fitting you in without busting minimum separation. Better to stick with your original routing, no matter how bad, and be assured of an approach slot when you arrive.

Tricking the system

There are lots of ways to get around the preferred route system, but not all of them work. Some can even blow up in your face, causing more of a delay than the preferred route would have in the first place.

There was one particular trip I remember: it should have been a routine flight. All we wanted was to go IFR from Bridgeport to Syracuse on a rainy Saturday night. When we called for clearance, the delivery controller sounded like he was one synapse away from a straight jacket.

"Look," he began, dispensing entirely with such niceties as the tail number or the clearance, "you can't do this." His anguished emphasis on the word *this* made him sound more psychotic than angry.

What had we done now? Nothing, other than to request RNAV direct from Bridgeport to Syracuse, a mere 180 miles. I knew it would never work. But I was in a hurry. The passengers arrived early and changed their destination at the last minute. Our charts were in the airplane so when the FSS guy asked for a route, we stuck in RNAV direct, figuring the host would pick up the city pairs and issue a preferred route.

For some reason, it didn't. I'd probably filed a Center altitude or maybe the thing just hiccupped but anyway, the controller apparently had a strip that cleared us RNAV direct to Syracuse. As George Bush might say, this would not stand.

"I can't give you this clearance. I'm gonna have to reroute you. Expect a ten-minute delay," the controller said. While we waited, I considered the bright side. On callback, the controller had recovered his composure and hissed the new clearance to us through clenched teeth, which is how most New York controllers normally talk. We were comforted by this return to the familiar. Further, the new clearance added 25 miles to the trip, a detour we had fully expected. Besides, we probably couldn't have found Syracuse with the RNAV anyway; don't get to use it much.

Backfires

This is just a minor example of how hurried attempts to short cut the preferred route system can backfire, substituting a longer delay for the short one you hoped to avoid in the first place. I can list instances which yielded even worse delays, the point being this: if you're going to be clever, be careful.

This is especially true of the change-the-destination-enroute ruse that Mike Busch describes in the next section, if used regularly in the northeast. It works most of the time around the New York area, such that the route to your pseudo destination results in an overall trip that's remotely nearer to direct than the preferred route would have been.

Subterfuge can get awfully involved, however. To beat the host computer's obsessive tendency toward city-paired routes, you might have to file two or three shorter segments which involve enroute climbs and descents at inconvenient times.

One particularly annoying city pair I've never managed to beat successfully is Hartford to Atlantic City. ACY is 160 miles southwest of HFD (230 degrees direct) yet the multi-engine preferred route takes you 30 miles southeast, over Groton then along V139 over the Atlantic. If you can't cut corners with vectors or RNAV to intersections, this route adds 40 miles to the trip.

One way to shorten the preferred is to file two segments, one to Westchester and another to Atlantic City, with an airborne pickup just before arriving at Westchester. The preferred route from HPN to ACY is nearly direct so overall, this trick should result in very near a great-circle route.

The last time I tried to beat it via subterfuge, I got two spins while the controller tried to work me into the flow. Eventually, I wound up getting vectored onto the preferred route anyway, which more than ate up any savings in time and fuel. My mistake was changing destination too soon but I had wanted to do it before the controller started vectoring me for the approach.

Subterfuge sometimes requires perfect timing and that comes only from knowledge of the local airspace. Even when you've got that knowledge and are armed with either the A/FD or a fat printout of preferred routes, there are still unforeseen gotchas.

Most of the TEC routes northeastbound out of ACY, for example, are limited to 5000 feet. Heading up toward Hartford or Westchester, they're fairly direct and hug the shoreline. If you happen to slip and file for 7000 or higher to pick up better winds, the host will give you a preferred route that's in Center airspace, far out over the water. The host doesn't care what route you filed nor does it know or care that no sane pilot of a single will accept the overwater leg. The system is defaulted almost entirely to altitudes and city pairs convenient to ATC. It's like a railroad mainline with no sidings.

No guarantees

I haven't found a foolproof method to manipulate the preferred route structure in the New York area.

My approach is to put the ball back into ATC's court. If I can't accept or don't want the preferred route, I figure it's reasonable to expect ATC to come up with something that will work for both of us and I phone the appropriate facility to ask for it. By following this strategy, I've learned a couple of useful lessons.

Lesson one is that more controllers than not take pride in devising shortcuts; most genuinely want to give pilots requested routes. Sure, ATC has its share of malcontents but the majority of controllers are quite creative in working around the system's rigidities.

Second, every ATC facility is different; each has its own in-house tricks, policies and procedures. This stuff isn't published in the A/FD nor will it be obvious from preferred route lists. What works at one facility, won't necessarily apply at another. You simply have to phone or visit and pose the right questions.

PVD to Philly, direct

I'm not about to suggest that this strategy works every time but when it works well, it works very well. On a flight last summer, I managed to negotiate—actually, it was offered—a VFR-on-top clearance direct from Providence to Philadelphia, right through the top of the New York TCA.

It was a miserably hot day, with the haze tops at 5000 feet. The airplane had no air-conditioning so I wanted to avoid the usual IFR hold downs and climb quickly to cooler air, on course if possible. I phoned clearance delivery and asked to work out a VFR-to-on-top clearance to a nearby VOR, with a cancellation at 6500 feet, then VFR to Philadelphia.

The controller explained that it wouldn't help, since all IFR departures were required to fly the local PDR, which was a bit roundabout. She asked if I could dodge the clouds well enough to accept a VFR climb. Sure, I said.

"Well, if we're gonna do that, might as well try for VFR-on-top all the way to Philly. Want me to work on it?" she said.

"Please," I replied, figuring there was no way that would ever happen.

"Okay, call me on 126.65 when you're ready and I'll have it."

True to her word, that's what she did. Our clearance was "cleared to Philadelphia, direct, maintain VFR-on-top." It worked too, until just southwest of New York, where the controller balked at handling us VFR-on-top. It didn't matter, though, because we had already shaved 10 minutes off the flight. We cancelled and went directly into Philly.

Unpublished preferred routes

Departing Wilmington, Del. a few weeks later, the trick worked again. I phoned the Philadelphia tracon and asked them to work out a clearance direct from Wilmington to Oxford, Conn. I wanted to avoid being shunted eastbound, then up V16, the preferred route.

"What altitude you want?" the controller asked.

"How about seven?" I replied.

"Nah, can't do it. McGuire won't take you at seven. How about five?" he countered.

"Okay, what route will work?"

He then proposed an oddball route over Northeast Philly to an unnamed intersection formed by two VOR radials, then direct on the airway. Except for a trivial jink, it was direct routing.

"Great," I said, "can you work that out with McGuire?"

"Don't have to work it out. It's a preferred route," he replied.

When I got home, I checked the A/FD carefully but that route wasn't published. A few weeks later, I visited the Philadelphia tracon to ask about it and sure enough, the route was among many that aren't published. Actually, according to Charlie Freeman and Rick Casey, who showed me around the tracon, the routes had been submitted for publication in time for the A/FD deadline but, for reasons unknown, they were left out of two issues running.

Conferring with clearance delivery has other advantages. If there's ice or unexpected delay or Flow Control is in the works, you'll hear the latest information from a credible source rather than second-hand through Flight Service or not at all through DUAT.

It's possible to file directly with clearance delivery, assuming the controller has the time to plug your flight plan directly into a FDIO. But unless you really need something special or have questions on routing, file the right way, via FSS or DUAT. After all, the clearance controller's job is to dig clearances out of the computer, not punch them in. Further, filing correctly keeps you in compliance with all the requirements listed in FAR 91.169, a legal fine point that might matter some day.

I like to file first then call clearance after the strip has printed out. Lately, I've just been asking for a better preferred route or something more direct. If it works, fine, if not, maybe next time.

It's surprising how many decent preferred routes exist that aren't published or, for whatever reason, just don't make it beyond the host computer. The only way to find out about them—short of carting around the printout—is to ask.

High-time CFII and 310 pilot Mike Busch adds his own strategies to the Quest For The Perfect Route, with a couple of ways to outwit theATC computer...sometimes.

Cloak-And-Dagger Route Planning

A lot of the IFR routes we get from ATC are positively obscene. Flying from point A to point B is rarely as simple as filing the most logical route. For reasons even the FAA doesn't always understand, clearances often add 50 percent to the trip mileage or they route us over hostile terrain with MEAs above the freezing level or they send us so far over water that a Grumman Widgeon pilot would need a change of underwear.

As we've noted before, it doesn't seem to matter one whit what we file or what pleas we add in the remarks box of the flight plan form. Sometimes it seems as though there's an FAA conspiracy intended to convince general aviation pilots to stay out of the IFR system. But that doesn't mean we have to accept this kind of treatment. For all its automation and rigidity, the ATC system is far from the impregnable monolith we imagine it to be. There are chinks in the armor. If you're willing to do a little studying and resort to some deviousness, it's possible to quietly slip through the cracks.

Know thine enemy

First of all, know your enemy. Don't get mad at the FSS specialist who copied your flight plan, or the Center flight data person who tore off your strip and reviewed it, or the clearance delivery controller who read you that horrendous route. Your true adversary isn't human. It's the ATC computer. To be precise, it's the host computer in the ARTCC where your flight plan originates. In most cases, automation has purged all traces of flexibility and compassion from the system and replaced these virtues with the heartless predictability of a computer program.

The computer looks at your airport of origin and destination and checks them against its database of "canned" routes called Preferential Departure and Arrival Route (PDARs), Preferential Departure Routes (PDRs), and Preferential Arrival Routes (PARs). It almost always finds an applicable P-route of some sort and that's what you'll get from clearance delivery. Guaranteed. The route you filed is often ignored altogether.

Man versus computer

I first became aware of how completely computers have taken over route selection a few years ago. I was visiting a pilot friend who was a controller at Oakland Center. I'd spent a few hours plugged in on-position, watching my buddy shepherd targets across his radar display. Finally, when it was time for me to fly home, I asked my friend if he could save me a call to FSS by entering a flight plan for me.

He said sure, then led me over to a nearby Flight Data Input Output terminal (called a "fido" in ATC-speak). There, he punched in my flight plan from Hayward (HWD) to Santa Maria (SMX). HWD is right in the heart of the congested Bay area airspace. SMX is 180 miles southeast of Hayward, out in the boondocks.

When the flight strip printed out, it gave the standard route: HWD..OAK..V244..ECA..V113.MQO..SMX. In case you don't have a chart handy, I can tell you that this route adds 60 miles to the direct route between Hayward and Santa Maria because it takes you over Manteca (ECA). I joked with my friend about what a terrible route this was. "Hang on," he said, "we ought to be able to do better than that."

He proceeded to try to coax a more direct route out of the computer. He would type in some undecipherable computer command, hit the enter key with a flourish, and exclaim: "There! That should do it." And each time, the new strip had precisely the same circuitous route as before.

He kept at it for 10 minutes, trying every trick he knew to override the computer's route assignment. Nothing worked.

Give up? Not yet

That's okay," I sighed, "I'll just have to live with the Manteca tour, as usual."

"Don't give up so easily," he said. "I know one sure-fire way to get a decent route." His scowl transformed into an evil grin. "We'll cheat."

He took a deep breath, cracked his knuckles, and attacked the FDIO once again. This time, he entered a flight plan from Hayward to Salinas (SNS). SNS is about 65 miles southeast of HWD and it's almost precisely on the great-circle route from HWD to SMX. The printer generated yet another flight strip, this time with a virtually direct IFR routing from HWD to SNS.

"Ha! Even the computer isn't cold-hearted enough to send you to Manteca if you're only going as far as Salinas," my friend said excitedly. "It would be nearly triple the distance."

"Hey, wait a minute," I protested, "I don't want to *land* at SNS. I'm going to SMX!"

"Not to worry," he said. "As soon as Bay tracon hands you off to Oakland Center, just tell the controller that you want to amend your destination to SMX via SNS.V25.PRB..SMX, which is virtually a straight shot."

"Won't the Center controller be upset if I throw him a curve like that?"

"Not at all. Happens all the time. It's no sweat for the controller...just a few keystrokes. Trust me."

"But, but...that's cheating, " I stammered.

"Sure it is," my controller-friend replied brightly. "It'll save you 20 minutes of flying time and fuel, and won't inconvenience anybody. It's a win-win situation. Ain't life wonderful?"

The Salinas subterfuge

A half-hour later I was starting engines on the ramp at HWD and calling for clearance to Santa Mar..er, Salinas. Cleared as filed, of course. I launched into IMC. Bay departure gave me a few vectors to thread me between the SFO and OAK arrivals, then turned me loose to fly direct SNS when able. When Bay told me to contact Oakland Center, I knew my moment of truth was near.

"Oakland Center, twin Cessna 2638X, seven thousand, request."

"Twin Cessna 2638X, Oakland Center, say request."

"Aaah, Oakland...twin Cessna 38X needs to amend our destination. Ahh... the...uh...boss just decided he wants to go to Santa Maria instead." I looked back at the empty cabin and verified that I was the sole occupant of the aircraft. "After Salinas, request Victor 25, Paso Robles, direct Santa Maria."

"Twin Cessna 2638X, cleared as requested, maintain seven thousand."

Obviously, it worked. To make a long story short, I started using the "Salinas subterfuge" scheme regularly and I've extrapolated the technique to similar situations. For instance, I discovered that to get a halfway decent routing from Monterey to Hayward, the thing to do is pre-file to San Jose then amend the destination to Hayward. It invariably works like a charm. Nowadays, I even leave out all the stammering and the reference to my fictitious fickle boss.

Bait-and-switch

Let's review the basic technique. You want to fly from A to B, but you know from past experience that the ATC computer insists on issuing a circuitous or otherwise objectionable preferred route. So you find an intermediate airport, such that the preferred routes from A to X and from X to B are both acceptable. Then you file from A to X, and as you get close to X you amend your destination to B.

If X is in an area of known-hostile controllers (the northeast, mainly) and you're worried that ATC might hassle you about switching destinations with no warning, here's an advanced variation on the basic subterfuge. Pre-file two IFR flight plans, one from A to X and the other (with an appropriately delayed P-time) from X to B.

Launch from A toward X. As you approach X, advise ATC that you won't need to land after all, and that you'd like to pick up your pre-filed second-leg IFR from X to B. Even the most hard-hearted controller won't have the nerve to tell you that you have to land at X anyway.

Of course, the key to the success lies in artfully choosing your bogus destination airport such that you can get an efficient IFR routing from A to X and from X to B. To do this consistently well, it helps to have a good knowledge of Center's preferred routes. If you fly in the northeast, this is not a big problem. A huge number of preferred routes are published in the A/FD and Jeppesen Airway Manual. Keep in mind, however, that not all of the routes are published so you'll sometimes have to do a little digging.

If you're flying in southern California, the entire PDAR system is published in the A/FD and Jeppesen manual in the form of SoCal Golf Routes. In one area I know of—northern California—PDARs aren't published anywhere. Officially, that is. But they are available if you're enterprising enough.

I called Oakland Center to ask for a listing of northern California's PDARs and was eventually connected with the Center's training department. The training folks at Oakland sent me a 28-page printout showing PDARs between every conceivable pair of northern California airports. They also provided a set of gorgeous blueprints (suitable for wall-mounting) that show exactly how all the airspace in northern California is parcelled up.

My experience with Oakland Center was not a fluke. I've dealt with the training department at Los Angeles Center and found them to be equally helpful. However, just who handles such a request will vary from Center to Center. In most Centers, the maps will come from the training department but the routes list may come from the Airspace and Procedures section. Whatever, just tell them what you want and you'll be switched to the right office.

Caveats

One note of caution: Don't try to use the Salinas subterfuge or any of its variants if your real destination is a major jetport for which Flow

Control is in effect. If you're flying to a flow-controlled destination, you're probably wise to accept the preferred route, even if it isn't as direct as you'd like.

The reason is that Flow Control makes it very difficult for ATC to accommodate special requests. If you try to pick up an IFR clearance to a flow-controlled airport while you are airborne, you'll probably be turned down and not all that politely. Even some airports that aren't flow controlled—Teterboro, N.J. and Westchester, N.Y., for example—may have a local rush that will get you stuck in a hold until things cool down a little. Not much you can do about that.

But all is not lost for the pilot-cum-grifter seeking to beat Flow Control delays. There's a slightly devious way around Flow Control and it works most of the time. The key is understanding how Flow Control is administered.

Suppose your destination is SFO. The Traffic Management Unit specialist at Oakland Center declares an official "acceptance rate" for SFO, based on weather, runway closures, and so forth. If all runways are open and the weather is good enough to permit visual approaches, SFO's acceptance rate is about 60 landings per hour.

When the stratus rolls in and approaches are necessary, the acceptance rate drops to 28 per hour. The TMU specialist monitors a computer-generated "metering list" of IFR arrivals, sequenced by ETA. For aircraft that aren't yet in the air, the ETAs are based on the proposed departure time and the filed estimated time enroute. If the projected SFO arrivals during a particular period of time don't exceed SFO's acceptance rate, no flow delays are necessary. When the projected arrivals do exceed the acceptance rate, the computer calculates an appropriate delay for each not-yet-airborne inbound flight such that the arrival rate is reduced to the required figure.

Beating flow delays

So how do you beat the system? Easy. Lie about your proposed departure time. Say you plan to depart for SFO at 1600Z to drop off a passenger departing SFO on an air carrier flight. By listening on your air-band scanner at home, you learn that Flow Control is in effect for SFO, and ground delays are 30 minutes.

If you were an honest and upstanding citizen-pilot, you'd file your IFR flight plan for 1530Z, leave for the airport 30 minutes early to allow time to complete pre-flight and checklists for your 1530Z P-time. Then you'd spend a half hour sitting in the runup area waiting for your Flow window.

But since you're out to beat the system, you file your P-time as 1530Z but leave for the airport in time for a 1600Z departure as originally intended. As you pull into the parking lot around 1530Z, you call clearance delivery on your handheld and copy your clearance and your Estimated Departure Clearance Time (EDCT) of 1600-ish. You then do a relaxed pre-flight and call for taxi at 1555Z, and finish your run-up just as your Flow window arrives.

This technique isn't foolproof, because you can't always guess at how much of a Flow delay you'll receive. Things change. If it turns out that the TMU assigns you an actual Flow delay of 45 minutes (from your P-time), you'll have shaved off 30 minutes of that and will sit in the aircraft for 15 minutes (still a good deal).

It doesn't pay to cut things too close when backing up your P-time to beat an anticipated Flow Control delay. If you guess that you'll have a 30-minute delay, maybe you should file for a P-time that's only 20 minutes earlier than the time you actually expect to be ready to depart. No point in trying to be too greedy. After all, the seasoned cheater is content to beat the system. He doesn't insist on pulverizing it.

Mike Busch just introduced us to another delay-causing procedure cooked up by ATC: Flow Control. It's supposed to make everything work better by avoiding aerial logjams, but it doesn't always work. O'Hare controller Denny Cunningham explains what it's supposed to do, and why it doesn't work.

Flow Control Failures

Okay, I'll admit it. The *idea* has an undeniable appeal: no stacks of holding airliners gobbling Jet-A as they await approach clearance; no Conga lines of departures stretching from one side of the airport and back again; no vectors for spacing, all of which adds up to the maximum use of airspace and airports.

This is what Flow Control is *supposed* to do by evening out peaks and valleys so there's a steady and orderly flow of airplanes, all choreographed to launch off their departure airports at predetermined times, fly standardized routes at precise intervals, arrive at the arrival fixes exactly on schedule, slide onto the final approach courses in a never ending rhythm, then deposit passengers and cargo at the gate with a minimum of delay and a maximum of efficiency.

Only problem is, it doesn't work. From where I sit in the tower, Flow Control isn't a dream—it's a nightmare. But before I tell you exactly why it doesn't work, let me explain how it's supposed to work.

Flow Control trickles into the system from two sources; a central office in Washington and from various ARTCCs. Whether originated from central or a Center, Flow takes the form of various programs intended to regulate traffic at busy airports. First and foremost are "in-trail restrictions," which means the longitudinal spacing that O'Hare (and others) provide between airplanes on the same route at the time of radar handoff to Chicago Center. While 5 miles in-trail is the legal minimum, 7 miles is the standard. During periods covering several hours each day, 10 miles is routine on some routes.

Another scheme is EDCT or Expected Departure Clearance Time. EDCT times are issued to airplanes bound for as many as a dozen different destinations, and these times are supposed to be adhered to within a five minute leeway, which is slightly more forgiving than the Enroute Spacing Program (ESP) to various airports.

ESP requires a phone call to the Center prior to release of an airplane to any participating airport, the idea being that the airplane will be granted a window of two or three minutes for release so that it fits into a high-altitude gap in arrivals destined for a particular airport.

Now, I'm not trying to cast aspersions at the fine folks at the Center, but I find it hard to believe that a two minute difference in departure times off the airport is going to make a significant difference when it comes to fitting a 500-knot airplane into a 20-mile wide gap located 5 miles above the earth— particularly since the Center doesn't know in what direction the airplane will depart or how much low altitude vectoring it will endure at the hands of O'Hare departure before it reaches the high altitude structure.

The three programs—in-trail, EDCT, and ESP—are used in combination, which makes things a tad complicated for a tower trying to manage a line of 30 or 40 departures, each of whom has anywhere from one to three different flow restrictions invoked prior to IFR release.

Ground controllers work hard to avoid putting airplanes filed for the same route back-to-back at the runway, so that airplanes in the back of the pack aren't delayed while airplanes in the front wait for the appropriate in-trail spacing.

Even if the ground controller has been 100 percent successful in this effort, his best laid plans are quickly put awry by the inclusion of numerous airplanes that Flow has decreed shall be subject to EDCTs, ESP or both. As the holding pads fill with departures awaiting their magic time, the departure line starts to backfill, sometimes engulfing arrival aircraft who desire nothing more than an unobstructed route to the general aviation ramp, which unfortunately lies adjacent to one of O'Hare's busiest taxiways.

Imagine the hopeless feeling of the three executives once seen relieving themselves alongside their Commander 580. After 45 minutes stuck in the departure line, their coffee intake conspired with the airport gridlock to make the prospect of reaching the general aviation terminal an unbearable option.

In case you haven't figured out why Flow Control doesn't work, let me elucidate. Flow Control is in a windowless room in FAA Headquarters and is so far removed from the action that every move they make is, by necessity, a reaction. It's like working a puppet on strings from the top of the Empire State Building— the puppet is so far away that by the time you move the strings, he's already turned in a different direction.

Thanks to Flow's stop-and-go tactics, we seen a phenomenon at O'Hare that was unheard of before FC came along: Absolutely empty final approach courses, for periods of 30 minutes to an hour, at several different times of any given day.

While the controllers are relaxing with coffee cups in hand, marveling at the fact that it's 4 p.m. on a Monday afternoon and there are only four airliners being vectored for the approach within a 40-mile radius of Chicago, thousands of passengers in hundreds of airliners all over the world are cooling their heels on airport taxiways.

They're waiting for their Expect Departure Clearance Time (EDCT) to arrive, under the misassumption that by absorbing the anticipated delay on the ground they'll arrive in Chicago without enroute delays or holding.

Of course, the piper must eventually be paid for those periods when arrivals have slowed to a dribble. When the airplanes start arriving in gaggles, approach starts vectoring, slowing, and holding, trying (as they always have) to make 10 pounds of guano fit into a 5-pound bag.

Last time I rode in an airline cockpit to O'Hare, the first thing the Captain did was ask Center for direct to an arrival gate for ORD, and a speed increase, both of which were approved.

The Captain turned to me and said: "We spent 20 minutes sitting in the runup pad waiting for the confounded EDCT. The direct routing and the speed increase will make up for most of that, so we'll arrive at O'Hare at about the same time as if we'd left on schedule. Some system we've got here."

The air traffic system is, by its very nature, a dynamic and constantly changing environment that needs innovation and creativity to operate at peak efficiency. Between wind, weather, and the various staffing and skill levels of air traffic facilities, there are too many variables for one central authority to manage the system effectively.

The fact is that Flow Control has become all too similar to weather, in that they are both predictable only in their unpredictability, and the vagaries of both are costly to all, in time and in money.

ATC's Limits, and How They Affect You

Controllers play by a special and separate set of rules that, while they don't apply to pilots, nevertheless affect us directly. Often a controller may direct us to do something for reasons we don't understand: chances are he or she is working under constraints we aren't aware of.

Knowing what drives a controller's actions can help pilots make their way more efficiently through the system. To that end, this chapter will deal with some of the rules, procedures and system features that compel controllers to do what they do.

First, Air Force F-16 pilot J. Ross Russo talks about ATC's need to keep airplanes widely separated from one another, even in good weather.

Keeping Everyone Apart

It's sometimes difficult for pilots to understand why air traffic controllers do the things they do. They often reject our requests for altitude changes, vector us away from our destinations, or occasionally request what seem like unnecessary speed adjustments. These delays are bad enough when IFR in real weather but they're downright aggravating when you're IFR in good weather. If it's clear and-a-million miles and there's not another soul in sight, trust me fellas, I'm not gonna hit anybody.

A pilot's impatience is understandable. After all, our aircraft are supposed to be an efficient means of transportation. Anytime we're deviated from the planned route or altitude, it costs time and money. Unfortunately, nobody ever seems to tell the pilot what's going on. Therein lies the most profound disconnect between pilot and controller, one that fosters mutual distrust.

It helps to know just what it is a controller is supposed to do. Simply stated, ATC's job is to "safely expedite the flow of air traffic." To achieve this, controllers must keep the airplanes in their charge pointed more or less in the direction they're supposed to be going and sufficiently separated to keep them from running into each other. Separation is the key word here because when the clouds are full of airplanes, pilots can't tend to separation on their own; they have to rely entirely on ATC.

To understand how separation works—and why it causes so much grief—it helps to learn a bit of FAAese. Strangely enough, what pilots know as "center" and "approach/departure control" and "tower" have different names in the FAA. The FAA refers to centers as "enroute" facilities. Approach/departure controls and towers are known collectively as "terminal" facilities. Elementary you say? Maybe, but not to the airline captain who recently filed a complaint to the FAA through his union.

The pilot, it seems, was instructed to use "visual separation standards" while working with an approach control. As would most of us, the pilot relied on AIM paragraph 274 for his definition of visual separation standards. Among other things, the AIM says visual separation is to be employed only in terminal areas.

To the greybeard in the left seat, "terminal" meant airport traffic area. Most pilots would make the same assumption. When the Air Line Pilot's Association followed up with the FAA, the feds insisted that proper procedures were being followed. Unfortunately, ALPA neglected to address the root cause of the misunderstanding, which is that pilots and controllers don't always speak the same language.

Money-back guarantee

Let's begin our discussion by reviewing some of the ways controllers keep us from running into each other. First of all, ATC's primary guarantee is a basic vertical and horizontal separation standard for IFR aircraft. Note the emphasis on IFR. In some cases, ATC is obligated to separate VFR traffic as well and there are instances when VFR aircraft are handled more expediently.

But in general, the separation guarantee extends only to IFR aircraft. These days, separation is is usually (though not always) achieved by radar so I'll assume we're flying in radar contact. Separation standards vary according to who owns the airspace you're flying in. Vertically, no IFR aircraft may come within 1,000 feet of another IFR aircraft, thus if you're cruising westbound at 6,000 feet, your protected airspace extends from 5,000 feet to 7,000 feet. Note that this is in accordance with FAR 91.121, IFR cruising altitudes in uncontrolled airspace. If you're in controlled airspace, which is usually the case, you have to fly at the altitude assigned by ATC. The minimum horizontal separation, on the other hand, depends solely upon your distance from the radar antenna. IFR aircraft within 40 nautical miles of the antenna site must be given a minimum horizontal separation of three nautical miles. Those flying beyond 40 miles from the antenna must be separated by five miles. In reality, this means that you'll be guaranteed five miles of separation when working with an enroute facility and three miles when working with a terminal facility.

These are minimum separation standards. The controller may throw in a mile or two of additional separation as a fudge factor so in practice, you're likely to get seven or eight miles from a center and five with approach. There's good reason for the extra spacing. A controller's life is made very unpleasant if he or she allows two airplanes to bust minimum separation. Center radars, that is enroute facilities, are equipped with a sinister device called a snitch patch. We'll look more closely at the snitch later. Essentially, the snitch watches over the controller's shoulder, sounding an alarm if two aircraft bust minimum separation.

Here's an example that puts the snitch into the pilot's perspective: Let's say your aircraft is equipped with an onboard computer that notifies your boss when you exceed a speed of 200 knots. Company policy is to ground pilots without pay for six months if they set off the alarm. The computer doesn't care if you're slow or arrive late at your destination, it only cares if you're over the 200-knot limit. How would you deal with it? I'm sure you'd do the same thing I'd do; fly 50 knots slow, just in case. If you picture a number of aircraft, each with different performance characteristics all converging on the same VOR and all asking for a lower altitude so they can set up for the approach, you can see why controllers get maxed out.

Tricking the snitch

Enroute separation standards, as monitored by the snitch, result in some truly ridiculous situations. Not too long ago, I was I was IFR northbound on V-225 near Key West. I was being held level at 4,000 feet by Miami center, even though I was anxious to get more altitude between me and the Gulf. The reason? Opposite direction IFR traffic. A Cherokee was southbound on V-225 at 5,000 feet, 12 miles ahead. I knew the controller couldn't give me higher because my climb would take me right through the Cherokee's cylinder of protected airspace.

Once the controller called the traffic passing off my left wing, I assumed I'd be cleared to climb. Not so. I had to wait until the Cherokee was at least five miles behind me. Climbing any sooner would have set off the snitch. Even if I'd seen the traffic, which I hadn't, it wouldn't have made any difference. Remember, visual separation isn't allowed in ARTCC airspace.

As general aviation pilots, we can exercise a couple of tricks to get around this situation which aren't available to our air carrier or military brethren. We can, in effect, put the IFR separation standards on hold temporarily and get treated like a VFR kinda guy. Perhaps the two most useful ways of doing this are the VFR climb or descent and the VFR-on-top clearance. The VFR climb or descent, if requested by the pilot and approved by the controller, allows you to operate VFR while you're changing your altitude. Once you reach your new altitude, you're IFR again. Of course, you have to meet VFR cloud clearance and visibility requirements during the climb or descent. Had I requested a VFR climb off Key West, I probably wouldn't have had a problem. However, because of increased VFR separation standards, this trick might not work in a TCA or a TRSA. A VFR-on-top clearance is similar. It allows you to operate VFR (at the appropriate hemispheric altitude), but also keeps you in the IFR system. Since you're still an IFR aircraft, controllers are required to provide you with traffic advisories and safety alerts in ARTCC airspace and separation in a terminal area. You get the best of both worlds. You're still IFR if you need to punch through some clag on the approach but your separation standards are greatly reduced.

But there are a couple of catches. First of all, pilots have to to ask for VFR-on-top. Controllers are specifically prohibited from suggesting it, which is as it should be. A really adventurous controller might say something like this: "Say again...understand you're requesting VFR-on-top?" Take the hint and you'll get what you want.

Because they aren't all that common, a controller may occasionally forget that VFR-on-top clearances are authorized. So if you've negotiated one with one controller and you're later instructed to contact a new controller "for further advisories," he may think you're VFR. Start asking questions.

Terminal talk

Once an aircraft enters the terminal area, controllers and pilots have more separation options. Besides smaller horizontal separation, controllers are allowed to authorize what's called diverging course separation. The concept is simple enough to grasp. If the controller observes on radar that two aircraft have passed and are on diverging courses, that's all the separation that's needed. He is therefore not required to issue vectors or altitude changes to maintain the three mile, 1,000 foot standard generally in force in the terminal area. Of course, that would never work with an ARTCC, even though it might be safe. The reason is that ARTCC gets its data from a mosaic of radar sites, each of which views the world a little differently.

It's possible, for example, for two targets to come from separate radars, even though the airplanes are right next to each other. So the targets may not have passed each other or diverged, even though the controller thinks they have.

Something called "conflict resolution" is allowed in ARSAs. Conflict resolution means the controller can see space between the two targets on his screen, "green between" in controller-speak or, as the pilot might hear it, "traffic no factor." This allows a controller to permit an IFR to pass within 500 feet of a VFR aircraft with no horizontal separation or no vertical separation with green between.

It's easy to see how much more efficient this is in a high-density terminal area and it's safe, as long as the visibility is good. In fact, conflict resolution is so useful that the the FAA allows the less-strict ARSA separation standards on a trial basis in several TCAs, including Los Angeles.

Cleared for the visual

As the airline pilot mentioned earlier found out, visual separation standards are allowed only in terminal areas. The requirement couldn't be simpler. If the pilot reports that he has traffic in sight, he may be instructed to "maintain visual separation with that traffic." In other words, get as close as you like but don't hit him.

Additionally, if a tower controller can see both aircraft, he may also use visual separation. The difference lies in who has responsibility for the separation. Obviously, if you report having your traffic in sight, you're responsible. That's why it's important to notify ATC should you lose sight of your traffic. This is no time to play "I've got a secret." If the tower is making the call, the controller is responsible for separation. He'll point out the traffic for you and, if you're in the pattern, he may offer to call your turn to base or final.

Visual separation standards go hand in hand with visual and contact approaches. The important thing to remember about a visual approach is that the airport can't be reporting weather that's less than VFR in order for a controller to authorize it. In a control zone, that's at least a 1,000 foot ceiling and three statute miles of visibility. At an uncontrolled field with no weather reporting, the pilot will be told "no weather available for Podunk muni" but the controller can base his approval on weather reports from nearby airports or from pilot reports. In order to get vectors to a visual approach, the ceiling must be at least 500 feet above the vectoring altitude (MVA) at an airport that reports weather.

If weather isn't available, you can still get vectors, providing the controller has some assurance that VFR conditions exist or if you've been told no weather is available. You'll usually have no way of knowing what the MVA is in the area you're in. The first clue will be the controller's denial of a request for vectors to the visual. Here's an example: You're approaching an airport from the west, where it's bounded by mountains. Another aircraft is approaching from the flatlands east of the airport. The other guy might get vectors to the visual but you won't because you're in an area where the MVA is higher, due to terrain. There will be circumstances when you'll be vectored for an instrument approach in marginal or good VFR weather. If you don't want the approach and can see the field, by all means report it in sight. You'll be immediately cleared for the visual, traffic permitting.

Believe it or not, there are instances when an airport is reporting IMC weather conditions when pilots are calling the field from ten miles out. Can you be cleared for the visual in this situation? No, even though you can see the field, the rules say no visual approach if the reported weather is less than VFR.

I've heard some very interesting exchanges between controllers and pilots in these circumstances. It usually happens when the field is reporting something like two-and-a-half miles in fog but really has inflight visibility of ten miles or better. Pilots will call the field and ask for the visual, only to be denied. Sometimes, though, the controller is pitching but the pilot's not catching: "Sir, I understand you have visual CONTACT with the runway, but the field is IFR, and I'm unable to approve a visual approach."

He's hinting at a contact approach. Unlike its "visual" cousin, the contact approach must be requested by the pilot. It can't be issued by a controller. To approve it, the controller has to be sure that the reported visibility at the airport is at least one mile and the pilot has to remain clear of clouds. And, of course, the airport has to have an approved instrument approach procedure.

Having the airport in sight is not required. According to AIM paragraph 402, the controller will provide "approved" separation. That means, among other things, that the controller won't assign a specific altitude but will clear the pilot for the approach at an altitude at least 1,000 feet below any other IFR traffic.

Runway in sight

I always thought that once I was cleared to land, I owned the entire runway. But that's not necessarily so. Just as there are separation standards for centers and approach/departure, so too are there standards for the runway. Runway spacing is based on three categories of aircraft. For simplicity's sake, we'll say that category I aircraft are propeller-driven singles, not including high performance aircraft such as the T-28. Category II aircraft are light, propeller-driven twins weighing 12,500 pounds or less. And category III aircraft are all others, up to jet transports. We'll just call them singles, twins, and large aircraft.

The basic runway separation standard for arrivals is simple: an arriving aircraft should not cross the threshold until the aircraft ahead of it has landed and taxied off the runway. However, during daylight if the visibility is good enough to estimate distance by using suitable landmarks, a controller has the option of compressing the spacing. Here are some examples: If a single is landing behind another single, only 3,000 feet of runway separation is required. So despite what your instructor may have told you about automatic go-arounds, it's okay to land with another airplane on the runway. If you don't want to, fine. Exercise your option to go around. If a twin is landing behind another twin or a single, 4,500 feet is required. And if either aircraft is large, 6,000 feet is required. At night or if distances can't be accurately determined, full runway separation will be used.

For departures, the basic standard calls for no takeoff clearance until the preceding aircraft has crossed the departure end of the runway or turned to avert any conflicts. When visibility allows, the same standards that apply for arrivals are used for departures, giving the controller the option of moving more metal on a busy day. However, at an airport with lots of mixed traffic, wake turbulence separation may slow things down some. Any aircraft departing on the same runway behind a heavy jet is supposed to be held for two minutes to allow the heavy's wake to dissipate. The departing pilot can ask for more separation or waive it entirely but in either case, he should make the request before taxiing onto the runway.

The snitch patch is just one of the gadgets the FAA uses to keep tabs on people. High-time CFII Mike Busch now discusses the features and limitations of the radar traffic control system that keep pilots and controllers honest.

The Computerized Big Brother

Picture this: You're flying a textbook VOR approach. ATC gave you a perfect radar vector to final and you're ahead of the airplane. Crossing the FAF, you start the timer, drop the gear, and pull the power back for a nice, steep 1,000 foot-per-minute non-precision descent. You've committed the MDA and the initial missed-approach procedure to memory. Any moment, you expect to break out of the overcast when suddenly, you hear this:

"Viking Five Nine Two, low-altitude alert! Check your altitude immediately. The MDA is 460 feet. Acknowledge."

The adrenalin rush gives you goose bumps and tachycardia. You check your altimeter: 600 feet and descending. The Kollsman window is set to 30.06 in., exactly what the ATIS said. The approach plate gives the MDA as 460 feet.

What's going on here? You've done everything by the book. You're not even down to MDA yet, much less below it. If you ask, the controller will give you the same explanation that you got from American Express when you called to complain about that $17 million charge on your monthly statement: "Don't blame me, it's the computer's fault."

Automated, sort of

Most pilots are vaguely aware that ATC's radar system is tied into various kinds of computers. Centers have what's called a host computer while terminal facilities have a system known as ARTS, for Automated Radar Terminal System.

By modern standards, a lot of this equipment is ancient and over the years, it has accreted various add-ons. Among these are radar software "patches" that are supposed to automatically alert controllers who are otherwise too busy to notice potentially unsafe conditions.

There are two basic flavors of safety alert: a low-altitude alert, warning of proximity to terrain or obstructions and a traffic alert, warning of proximity to other aircraft.

These software patches are known to controllers by their alphabet-soup acronyms: CA means Conflict Alert, LAAS means Low-Altitude Alert System, MSAW means Minimum Safe Altitude Warning, and MCI means Mode-C Intruder. The most infamous software patch of all—the Operational Error Detection Patch—surely deserves a fancy acronym like OEDIPUS or somesuch, but instead, it's universally referred to by its disparaging nickname: snitch.

On balance, these automated alarms have been helpful in preventing accidents. But they've created their own set of new problems, namely nuisance alerts and a tendency for controllers to rely on them more than they should.

How low can you go?

Let's look at one of the earliest safety alerts first, the LAAS described earlier. LAAS still exists in a few low-rent approach control facilities that are equipped with ancient AN/TPX-42A radar gear. Under LAAS, the terminal area is divided into a grid and each square is assigned a minimum safe altitude based on known obstructions.

When the computer sees an IFR aircraft reporting a Mode-C altitude below the minimum safe altitude of its grid square, the computer sounds an alarm and highlights the target on the controller's radar scope. Sounds simple and logical, doesn't it?

In practice, LAAS turned out to have a number of deficiencies. It's quite common for Mode-C data to be momentarily corrupted by a phenomenon called "synchronous garble." This can occur when several aircraft are at a similar bearing and distance from the radar site. Garbled Mode-C data can cause LAAS to issue bogus low-altitude alerts.

Further, the minimum safe altitude threshold in each grid square can't simply be set to the published MEA, MOCA, MVA, MIA, or MDA. If it were, there would be a veritable tidal wave of nuisance alerts. Aircraft cruising at the MEA commonly make excursions of 100 or 200 feet from the assigned altitude. Altimetry problems and encoder slop can add another 200 feet of error to the Mode-C data. Thus, the LAAS grid thresholds must be set quite low to keep the incidence of nuisance alerts to an acceptable minimum.

This creates yet another problem. Pressurized aircraft normally descend at several thousand feet per minute. An alert triggered by a quick descent through the warning threshold doesn't provide much warning before terrain impact. It doesn't take much imagination to understand what could happen: "Ah...Mirage 648, low (splat!) altitude alert...oops, now where'd I put that clipboard full of accident/incident reporting forms?"

Worse yet, LAAS software is working on altitude data that's typically 6 to 12 seconds old, due to the rotating antenna and the limited computer processing speed. By virtue of its short warning time and its proclivity for nuisance alerts, simple-minded LAAS is sometimes more trouble that it's worth.

Tricky tracks

Today, most tracons use ARTS, a fairly sophisticated computer that maintains a track on each aircraft. The computer tracks all IFR beacon targets plus those VFRs who are receiving radar service.

For each tracked target, the computer maintains a history of the location and altitude readout of that target during each of the past several radar sweeps.

From this history, the computer calculates the target's course, horizontal and vertical velocity. It then extrapolates from this information and tries to predict where the target will be on the next radar sweep.

The ARTS' tracking ability made possible a greatly improved system called Minimum Safe Altitude Warning. MSAW eliminates most of the problems with LAAS. For one thing, garbled Mode-C returns aren't a problem because ARTS can correlate each Mode-C reply with prior altitude data from the track's history. It can thereby detect and ignore bogus Mode-C replies, eliminating a major source of nuisance alerts.

More important, track history allows MSAW to *predict* what the position and altitude of a tracked target will be, say, 30 seconds from now. The computer can generate an alert if the *predicted* altitude of the aircraft is below the minimum safe altitude for its *predicted* position. MSAW can issue alerts with significantly greater warning time than LAAS. The objective is to make sure that MSAW sounds its alarm early enough to give the controller time to transmit all of the required phraseology before the airplane splats. And MSAW usually does.

Not too smart

If MSAW has a weakness, it's that the computer knows nothing about clearances or pilot intentions. All it knows is what it sees on the radar scope. When MSAW sees an aircraft descending rapidly toward the MEA/MVA/MDA, it assumes that the aircraft will continue descending at that rate and consequently, it generates an alert. Such nuisance alerts are particularly common during non-precision approaches.

Now imagine for a moment that you are a radar controller. You see an MSAW alert flashing on the data block associated with one of the aircraft you are working. You know the pilot is making a non-precision approach, and so you're pretty sure that the alert is a false alarm.

What do you do? Do you issue a low altitude alert to the pilot, or do you disregard it? Theoretically (according to 7110.65, the ATC manual), it's your judgment call. But in practice, you have no real choice. If you don't issue the alert and the the aircraft winds up pranging, guess what happens to your career?

To reduce nuisance alerts, each tracon defines the airspace in the vicinity of busy airports as Type I areas, and the commonly used approach corridors as Type II areas. Everything else is considered Type III. MSAW is programmed to disable alerts in Type I areas, and to shorten its look-ahead time in Type II areas.

In addition, MSAW must detect a low-altitude problem with a target for two consecutive sweeps before it will generate an alert. This helps reduce the frequency of nuisance alerts due to momentary altitude excursions.

Overall, MSAW has received such high marks in tracons that the FAA installed a similar software enhancement known as Enroute Minimum Safe Altitude Warning (E-MSAW) at all Centers. So if you're in radar contact and squawking a discrete IFR code, it's almost a sure bet that MSAW is watching out for you.

MSAW can also watch tracked VFR targets that are squawking a discrete code. However, most ATC facilities disable MSAW on code blocks assigned to VFRs to reduce nuisance alerts.

Conflict alert

Just as it can use track data to issue terrain warnings, so too can ARTS (and the Center's host) warn of an impending collision or a separation bust. In a tracon, the Conflict Alert warning objective is 30 seconds; in a Center, it is 80 seconds.

In the Center, a CA-generated alert causes the data blocks of both aircraft to blink on the controller's radar display but there's no audible alarm. In a tracon, it puts up a blinking CA symbol in the upper righthand corner of each data block, and sounds an audible alarm as well, which most of us have heard in the background from time to time. When the controller sees the CA alert, he is supposed to warn the pilot(s) using the following phraseology: "Viking Fine Niner Two, traffic alert, one o'clock, less than a mile. Advise you turn left to 360 and climb to 5000 feet immediately."

Despite the sophistication of its software, CA still generates lots of nuisance alerts. Like MSAW, CA's Achilles' heel is that it doesn't know anything about clearances or pilot intentions. Imagine two IFR aircraft at the same altitude and on converging courses. CA predicts that the two targets will merge in 30 seconds and generates an alert.

The controller knows that there's really no problem because one of the aircraft is about to intercept the localizer and make a 30-degree left turn. Or perhaps one of the aircraft has reported the other one in sight and is maintaining visual separation.

So the controller ignores the computer alert and you never hear about it. Unfortunately, CA cries wolf so often (especially in a terminal area) that controllers are often tempted to disregard its warnings. Until the advent of datalink ATC, we can't do much about that.

CA works only on tracked targets. In a tracon, the computer usually tracks only transponder-equipped aircraft that are squawking a discrete code. This typically includes all IFRs and those VFRs that have requested radar service.

This means that if you are flying IFR and are about to be boresighted by a VFR aircraft squawking 1200, you had better hope the controller notices it; the computer definitely won't. This is one of the motivations behind TCAs and ARSAs; all aircraft in these areas (legally) will be tracked.

Flight of the intruder

Center radars are equipped with two other CA-based warning systems that tracons don't have. One is called MCI or Mode-C intruder. What MCI does is simply this: it looks for any Mode-C target that's squawking 1200 within Center-controlled airspace. If it sees an aircraft, it automatically starts a track on it, even though no discrete code has been issued.

Intruder tracks are identified on the Center controller's radar scope with the letter "I" and a limited data block that identifies the target as VFR and displays its altitude. Intruder tracks are processed by CA in much the same fashion as the tracked targets of IFR and participating VFR aircraft.

Pilots flying IFR in VMC assume that a controller sees every little blip on his scope in time to issue an advisory. The targets probably are there, but they can be overlooked. To help out in this situation, Center controllers have MCI, which is an enhancement of the conflict alert computer program.

Here's how it works. Suppose you're flying VFR at 11,500 feet and it's CAVU. Thirty-five miles off your right wing is Hubtown International. The last controller you worked terminated radar service and didn't offer a handoff but you figure, "Heck, I'm squawking 1200 with mode-C, so ATC can certainly see me."

Sitting in a dark room, a hundred or so miles away, the Center controller *does* see you. In fact, he notices your target about to cross his jet arrival corridor into Hubtown International. He's concerned because these jets will be sequenced to descend to 10,000 feet in your vicinity. His workload intensifies as he works out the sequence.

Number one is an easy choice today. It's a 727 leading the pack, 40 miles ahead of the closest company jet. A quick clearance instructs the pilot to cross 30 miles from Hubtown at 10,000 feet. The controller takes note of your target, which could be in conflict with this jet, but it's much too early to be certain.

The controller now turns his attention to sequencing numbers two, three, four and so on. Unexpectedly, numbers four and five deviate around building Cu. The controller's well-laid plans are trashed; a new strategy needs to be formulated.

Several amended clearances later, he glances back at the 727's target just as its pilot comes on the frequency: "Center, did you see that aircraft?" The controller, feeling somewhat embarrassed, realizes that he didn't follow up on the potential conflict between the 727 and the VFR target soon enough.

In ARTCC airspace, such a scenario is less likely to occur. MCI would have directed the controller's attention back to the 727 and the VFR target in time for a traffic advisory to be issued. As noted above, MCI is part of an expanded conflict alert system. It's set up to alert the controller when an aircraft, which is not being worked by anyone in the ARTCC, comes into conflict with an aircraft being provided Center radar services.

Its parameters are similar to the normal conflict alert; it flashes the datablocks if the targets will pass within close proximity within a certain minimum vertical separation. The alert begins about two minutes before a potential conflict. Although the dreaded snitch patch is also related to conflict alert, MCI has nothing to do with the snitch.

MCI is adapted only at ARTCCs, works only at altitudes of 5000 feet and above, and will not work unless both aircraft are mode-C equipped.

The bottom line on MCI is this: When you're flying IFR (or VFR with advisories) in Center airspace, you are now significantly more likely to receive traffic alerts on non-participating VFR aircraft than in the past. And that's the good news.

What isn't such good news (at least for controllers) is the notorious snitch patch. Again, it's found only in Centers and the idea behind it is simple. Whenever a loss of standard separation occurs between IFR aircraft, the computer alerts the Area Manager by sounding an alarm at his desk and printing out a report. The AM then trots over to the radar console where the possible loss of separation occurred and has a little talk with the controller to find out what went wrong.

There are three possibilities: 1) It's nobody's fault. A technical glitch set off the snitch. No foul. 2) It's the controller's fault, in which case he or she is immediately pulled off position and perhaps decertified pending remedial training. 3) It's the pilot's fault. If the controller succeeds in persuading the AM that he didn't screw up, guess who's gonna get a letter from the Friendly Feds?

The snitch and the FAA's it's-gotta-be-someone's-fault policy has had a chilling effect on the relationship between pilots and controllers during the last decade. Since the advent of snitch, Center controllers use six or seven mile spacing instead of the required five, just to be on the safe side. Can you blame them? As a result, Center controllers use more airspace than they used to and that produces more delays for all of us. It also means that we pilots are subject to a lot more violations and certificate actions than in the pre-snitch days.

Most pilots think that the moment they stray more than a certain amount from their assigned altitude they'll be automatically busted: that there's a gadget somewhere at ATC that keeps track of these things, and so fly in fear of losing their flying priveleges through a mistake.

It doesn't work quite that way, though as we've seen there are a variety of methods by which the ATC computers keep tabs on pilots. In the real world, an altitude deviation may go entirely unnoticed.

If it does get picked up, it may make no difference to anyone and nothing will come of it. The pilot, of course, may actually get busted. Or, as Denny Cunningham wrote in the "In the Hot Seat" chapter, it may result in a "deal" and the controller might feel the heat.

In this section, controller Paul Berge discusses the ins and outs of straying off altitude, and keeping yourself and the controller out of trouble.

Altitude Busts

"Cherokee Seven Four Whiskey, stop altitude squawk, altitude differs by 300 feet." Ho-hum, another flaky Mode-C trace. On an average day shift, it's not unusual to see a couple of these before lunch and until early 1992, no one really cared about them very much. Maybe no one cares much even now, but ATC has started keeping tabs on all bum Mode-C transponders.

We're supposed to copy the tail number and forward a weekly list to the local FSDO. What the FSDOs intend to do with this information is a little vague at the moment but it appears that the concern here is to weed out erroneous altitude readouts that might cause faulty TCAS alerts. The procedure is not meeting with universal compliance, but that doesn't mean it's being dropped. Someday, your "I'll get the Mode-C fixed next week" attitude might draw official attention. The process is just slow.

Noble intentions aside, a recurring question arises at Operation Raincheck pilot meetings where I occasionally speak: How much does ATC monitor a pilot's altitude, and what happens when a few hundred feet are accidentally misplaced? In other words, how paranoid should you really be? As is usually the case in these delicate matters, the answer is: It depends.

Justifiable fear

In the not-so-perfect world of ATC, there's an expression: "80 percent is passing." Don't reach for your AIM, it's not there. Found only in the FAA's secret 7210.3J, "Air Traffic Controller's Guide To The Universe," this fudge margin is the amount of leeway a controller has—if no one's looking—for squeezing that extra target onto the scope. Not really a license to use less than standard IFR separation, this 80 percent rule is more of a ranking system.

If a controller loses separation between two aircraft, say less than 1000 feet vertical, the controller is said to have had an operational error, an OE or, as Denny Cunningham explained in the "In the Hot Seat" chapter, a "deal." Under the 80 percent rule, there are deals and there are big deals. A big deal would be putting two aircraft at the same altitude and same place, so the pilots can wave to each other as they pass. ("Isn't that man wearing a funny tie!") This is the type deal that makes the papers and keeps the drug-testing companies in business.

When separation has only been compromised a tad, then the controllers is said to have had, "...no big deal." If 80 percent of required separation is intact, the controller can get off with a stern lecture. The local facility manager still investigates the incident, but lacking public outcry, the offending controller is not decertified and the whole affair is simply not mentioned in polite company.

Pilot error, too

And what's good for the goose is good for the gander, in this case an instrument-rated pilot who strays a few hundred feet from assigned altitude. Or is there a double standard?

FAR 91.123 (a) is clear: "When an ATC clearance has been obtained, no pilot in command may deviate from that clearance..." So "maintain 3000" means exactly 3000 feet and not a notch otherwise, unless there's turbulence, or if you drop a pencil and dip a little below 3000 feet, or maybe a climb was left unchecked until a few hundred extra feet whizzed past. No big deal.

Experienced controllers understand these slight deviations and turn a blind eye or give the pilot a gentle reminder to return to the assigned altitude. After all, ATC isn't one-sided; it's a cooperative endeavor wherein controller and pilot share a common goal: *not setting off the snitch.*

Developed by former East German Stasi technicians, the snitch continually monitors radar data for altitude deviations and expired medicals. Whenever two targets are within three minutes of violating each other's airspace, the Center data tags will flash to warn the controller of a potential loss of separation. Should this warning go unheeded and a separation bust result, the watch desk will receive a printout requiring the watch supervisor to investigate.

Did you think the snitch only applies in Center airspace? Nope, it works in some—but not all—approach airspace, too, in areas where Center radar also covers or overlaps approach airspace. The snitch is patched onto Center's computer (not approach's ARTS) so the Center supe is responsible for notifying the appropriate approach facility of a snitch-detected separation bust.

Usually, this is as far as it goes, as the facility responsible for the alarm will have some sort of excuse such as visual separation was being used or one of the aircraft had already canceled IFR. In either case, the quick investigation is recorded and filed should something come up later, such as a pilot calling to ask why he saw what looked strangely like a DC-10 blow past him in the clouds.

Two to tango

The key to altitude deviations is the presence of another aircraft. If you're all alone in your Mooney over North Dakota and you've accidentally drifted a few hundred feet from your assigned, you may receive a call from Center asking you to "verify altitude."

Before you speak, think. You have three choices at this point. 1) Lie. Turn off your Mode-C and tell the controller you're level at whatever your assigned altitude was (not recommended, although it works) or 2) confess: "Ooops, I'll hustle back up there. Sorry!" Or 3), feign comm loss, hustle to your correct altitude and then report being at the assigned. I'll let you decide which strategy to follow.

If there's no separation loss, the controller will probably grin and forget about it. Even the snitch will ignore it. However, if your deviation took you into another aircraft's path, both snitch and controller might not be smiling. If you've violated someone else's airspace, killing the Mode-C will probably just get you into deeper trouble.

A few years ago, a controller from Monterey climbed a King Air into Oakland Center's airspace without permission. Realizing his error, he panicked and told the King Air pilot to stop squawk. The pilot, becoming suspicious, wanted to know why and left his transponder on. By then it didn't matter, because Center had called, inquiring about the target they saw climbing through their traffic. So it's not just pilots who get busted.

And it's not only GA pilots who screw up, either. Before Republic Airlines joined the list of dead air carriers, I had the pleasure of watching a true pro covering his tracks. The Convair had been handed off to me from Minneapolis Center. He had been assigned 11,000 feet. Center shipped the pilot but before he checked on my frequency, I watched his altitude readout drop below 11,000 feet, then 10,000 feet.

Finally he checked in "...level at..." and silence. The Convair's Mode-C mysteriously vanished from the scope. Several calls to the pilot proved fruitless. Finally, the Republic pilot checked in again, this time "level one, one thousand," with the Mode-C miraculously back in working order.

He knew and I knew. And he knew I knew. But it didn't matter. The Convair pilot may have de facto busted an assigned altitude but in the absence of controller follow-up or a separation bust, he walked. It's the "If-a-tree-falls-in-the-forest-theory." If no one complains about a deviation within 15 days of the event, it never happened. Recorded tapes of all ATC radio transmissions are saved for 15 days just in case someone has a gripe. After that, everyone's clean.

Controllers aren't cops

So, what about all these FARs and stuff that say pilots must maintain assigned altitude? What constitutes an altitude bust? Who's the judge?

Let's tackle the last question first. If you've deviated from an assigned altitude and find yourself in the courtroom stage of the enforcement process, quoting anything from this article might only add to your sentence. Just let your attorney handle the boot-licking.

The pilot's first inkling that an ATC violation is in the mill begins with the controller. I once ran a stop sign (late for work) only to pass a semi-napping California Highway Patrolman not ten feet away. My case never reached the courts; I threw myself on the mercy of the bemused officer and took my lumps.

The same principle applies to ATC. At pilot meetings, I've explained: "We are controllers; not cops. We leave enforcement to someone else." Controllers have the luxury of pointing to FSDO and saying, "It's them! *They* do the dirty work!"

That used to hold true, somewhat. Unfortunately, the atmosphere has been undergoing subtle changes and a whole new generation of controller is being hatched. Where once a pilot could be bullied to "call the tower" for an unofficial telephone reprimand, the same request is, in some cases, handled in a nastier fashion.

Local directives of dubious origin may not only encourage controllers to become enforcers but often require us to violate that pilot/controller trust. In the Central Region, for example, any time any aircraft departs with the RVR below 1600 feet, the callsign is forwarded to FSDO. Regardless of whether the pilot is operating under Part 91, 121 or 135, there's a certain sniffing for violations going on.

The same applies to arrivals. Of course, Part 91 pilots are quickly absolved. Still, the attitude is fostered that we controllers are expected to snitch, snitch, snitch. Between you and me, we don't follow this mandate with great zeal nor do these kinds of policies exist throughout the system.

Contrary to pilot-lounge belief, FSDO usually knows nothing about airborne violations until someone from ATC reports one. Simply because a pilot is reported by ATC does not mean a controller, per se, did the snitching. A controller gains nothing by busting a pilot. A supervisor, however, may reap more political hay by being seen from above as a real "company man."

So you've busted an altitude, what then? Under the "Pilot Deviations" section of the 7210.3J, ATC is directed to report the date and time of an incident, the aircraft identification and pilot's name and address (if known) to FSDO within four hours of occurrence.

ATC is required to report pilots "When it appears that the actions of a pilot violate a Federal Aviation Regulation or NORAD ADIZ tolerance." A rule of thumb gives the pilot 300 feet of leeway for altitude. Any deviation over that rings the bell.

Once the pilot deviation reporting process begins, ATC has the option, work load permitting, to inform the pilot: "Cherokee Seven Four Whiskey, possible pilot deviation, advise you contact (facility) at (telephone number)." Don't think that wouldn't get the whole class's attention.

With the investigation process in motion, the facility manager refers to Order 8020, with associated forms, and a case is built against the deviant pilot before he ever reaches his destination. The FAA is very good at this.

Looking the other way

Disclaimer: Not all ATC facilities are created equal so not all follow up a potential violation with the same enthusiasm. This varies from facility to facility and from controller to controller.

A few months ago, a Cherokee departed one of our satellite airports, climbing to an assigned altitude of 5000 feet. I had no other traffic in his area and was vectoring someone to the ILS 25 miles away. When I looked back at the Cherokee, I noticed his altitude was 5600 feet and climbing.

"Cherokee 123, say altitude."

The pilot replied: "...we're out of four thousand six hundred for five." I was about to give the pilot the altimeter setting and ask him to double check everything when he keyed the mic and shouted: "Ah, ah, no...we're at five thousand, six hundred, ah, ah, ah...should we descend?"

As calmly as possible, I told the pilot to descend to 5000 feet. He complied and did what every pilot should do in that situation: he panicked. He asked me flat out if he was in trouble.

Beneficently, I explained, "Oh, no. It happens all the time; no separation was lost." I added a kindly lecture on watching altitude and assigned the pilot 30 Hail Marys to be said before he could descend again. That was the end of the incident. Or was it?

Technically, the controller is required to report *any* violation. The 7210.3J, paragraph 5-62 says so. I should have immediately called the supervisor away from the sports page and reported the incident. Then I'd be required to write a report and the supervisor would be required to call a string of hitmen on up the line. FSDO would be included in the list.

By electing not to officially destroy the pilot, I stuck my neck out slightly. My excuse was that I took care of the incident and safety was never compromised. Had there been other traffic, it might have been another story.

Personally, I like to think that the controller who ignores the letter of the regulations and doesn't automatically turn in a pilot deviation is exercising professional judgment. Both controller and pilot have too much to sweat when the ceiling's a ragged layer of mist and ice without wondering what paragraph of Washington regulation minutiae has been violated.

1984 is here

But what about the controller seated beside me? Who can you trust? That same 7210.3J requires *any employee* witnessing any violation anywhere to report the violation to the proper authorities. This Orwellian approach to safety keeps everyone looking over shoulders. What one controller might overlook, a coworker might report. Very few controllers take this seriously, although it only takes one to ruin your day.

If you're on good terms with your local tower, great; I hope it never changes. But the good ol' boys you've talked to for 20 years, the same ones you'd see at the EAA fly-ins and local hangar barbecues may be close to retirement. The replacements won't know you. Good ol' controller Fred might have been an instrument-rated pilot who loved flying as much as you, but his replacement may have never seen an airplane until hiring on.

Is this bad? Not always; just new. The future is populated with air traffic controllers who may never even sit inside a piston aircraft. In short, there's an attitude change on the way and the individual in the system is being replaced with the system itself. Where once we could exercise common sense and good judgment, we are more and more delving into manuals searching for official direction and precedent. Which is to say this: Be careful out there. don't assume anything.

We've seen how the system works to keep everyone, pilots and controllers alike, in line. But there are still some things that controllers do for apparently mysterious reasons.

For example, ever wonder why it sometimes seems as though you're always "low man on the totem pole?" It's often as though everyone else has priority over you. After all, it's supposed to be firt come, first served, and you have every bit as much right to be in that airspace as some fancy corporate pilot, right?

Well, surprise. In a busy terminal environment, might really does make right. Controllers will let the big boys in first, and for some very good reasons. Controller Paul Berge explains why.

No, After *You*

Growing up on the farm, each night we'd bring the old plow horse into the barn. At the same time, all the chickens, who'd been out eating bugs all day, would also return to roost. The problem was, our barn door was exceedingly narrow, limiting access. Each night I was forced to make the decision: who first? Big horse or little chicks? Each night, the 2000-pound hay burner would lumber through the doorway as chickens scattered and were occasionally squashed. Get the analogy?

Okay, I didn't grow up on a farm. In fact, I grew up near Teterboro, watching little airplanes getting squashed by big ones destined for Newark and LaGuardia. What I observed early on has carried through my professional life as a controller: "Run the big guys first, 'cause they make the most noise!"

Secret's out. The guy flying the airplane with the big motors and hundreds of paying customers on board tends to ease out the guy whose plane sounds like a very large housefly.

So, you're in your Mooney. You've filed IFR; followed all the rules and arrived as per your ETA. You check on the frequency with the ATIS (something those guys in the kerosene hogs don't always do), and you think you should be number one because you're closer to the airport and because you were there first.

It doesn't necessarily work that way. The first come, first served theory dissolves in the harsh realities of sequencing airplanes in the terminal environment. Ideally, all jets would land at one airport and pay huge landing fees and all GA planes would land at more convenient facilities, where the pilots would be welcomed by grinning FAA personnel offering to help. Reality dictates otherwise.

How fast do you think you are?

A Lear 35 dropping out of the flight levels, dips through 10,000 feet and "slows" to 250 knots. A Mooney cuts airspace at about half that rate.

Who's first? Roughly speaking, if both aircraft maintain a similar speed disparity, the Mooney on a 10-mile final blocks the localizer for 10 miles behind. It's easy to see, then, that a slow GA airplane blundering into the afternoon arrival push of 15 air carriers is a real problem for a controller. If all aircraft were Mooneys (eek!), we could run three miles in trail all week and never flinch.

But along comes the Lear asking, "I'm slowing to follow what?" The young controller, being afraid of pilots who sound like Charleton Heston, often panics and pulls the slower Mooney away from the final to slide in behind the jet. That's one excuse.

What about the more seasoned controller, savvy to the ways of speed control and eye-balling a slot for a screaming jet? The last thing an approach controller wants is a slow aircraft strung way out on the final, grinding along at 80 knots, 10 miles outside the marker.

Experience suggests that it's easier to run the faster aircraft in first and gently vector the slower one around until a slot is available. This is sometimes referred to as "boxing" the little guy. Pilots have other terms for it.

It works like this: You, in your Mooney, are vectored toward the final approach—or airport, if a visual approach—then, as you say "airport in sight," we say, "roger, turn right heading...," and take you completely away from your intended destination to follow some Chapter 11 airliner on a 15-mile final. You've been boxed; penalty boxed. You can cancel if you want but it won't do any good.

Why? You suspect it's because the controller owns stock in the airline (an illegal practice, by the way), but more likely, you just didn't fit. The approach controller must judge your ground speed and the jet's speed and guess who will arrive at the localizer (or runway if visuals are in use) first.

It's annoying to be told to follow someone miles to your right, when the airport is two miles to your left. The 737 (aka Guppy) you've been told to follow, however, may be covering a great deal more territory than you per second, plus the controller must consider how much you may slow inside the FAF versus how much Belly-Up Airlines slows on final.

Here are some handy things to remember: 1) Airlines in financial trouble always fly faster finals than airlines with sound balance sheets. Each morning, the supe checks the stock market reports to see how the respective airlines will be flying that day. This is called "rough-cut sequencing" or "handicapping." 2) Military aircraft (combat types) are always "minimum fuel" and will be sequenced ahead of all others. Plus, they carry guns. 3) Twins, except Apaches and Citations, always beat singles. 4) Turbines beat pistons; again, except Citations. 5) Mooneys are usually last. 6) Cessnas follow Mooneys.

More than a speed problem

The speed control lesson is constantly forgotten and is often relearned by controllers. Frankly, we sometimes screw up the sequencing. With ten targets on the scope, we make snap decisions to put the jets first, when more careful review would warrant running a single-engine airplane that's already closer to the field.

Too often, a bad sequence call only becomes apparent after it's made. Then everyone must live with the decision, and someone burns extra fuel. Plan for it; call it a penalty reserve. Of course, if you know about such fine points as approach gates, you can query the controller for a better slot.

Gates are aiming points for vectoring aircraft to a final approach course. The Pilot/Controller Glossary puts the gate ".1 mile from the outer marker... on the side away from the airport..." This "imaginary point," as the Glossary says, "will be no closer than 5 miles from the landing threshold."

To controllers, the gate has some importance. According to the ATC manual, controllers are supposed to vector arriving aircraft to intercept the final approach course at least 2 miles from outside the approach gate unless one of the following exists: Weather conditions better than a ceiling 500 feet above the MVA/MIA and visibility at least 3 miles, or if a pilot requests a tighter turn-on. In either case, the controller may *never* vector to intercept closer than the FAF.

Clearly, a vector inside the final approach fix is worthless but why should a pilot flying a 70-knot final be forced to fly three miles away from the marker, then three miles back for a total of six flying miles before ever reaching the marker? Because the rules say so, that's why.

As with contact approaches and special VFR clearances, it takes a pilot-initiated request to shorten up an approach. Should the pilot request "a turn at the marker" or some other shortcut, I'm happy to oblige. But, there are limits. Asking a controller to vector you to intercept the final at a specific point—say, the marker or a mile from the marker—is like a tower controller asking a pilot to "put it on the numbers." Some controllers might be good enough to nail a target to the final wherever they want but, frankly, I'm not one of them. Some days, I'm lucky to hit the localizer without issuing apologies.

Ask for a tight turn soon enough to give the controller time to plan a little Kentucky windage. You may even find yourself moving up from number seven in line to number one. Then again, your request may also fall on deaf ears. Regardless of groundspeed, trying to squeeze a target from a downwind to a slot in front of someone established on final eats up a lot of territory. What might appear to be a tempting opening may turn out to be a DAT (Dead Ass Tie).

One bit of advice: Don't develop a reputation with ATC as "that guy who's always trying to call the sequence himself." You may think you have the ATC picture more clearly than the befuddled controller below, but don't succumb. Don't become a pest with a better plan.

800-pound gorillas

Most controllers know that it's easier to move something small out of the path of something big. Physical size gets results. I once stepped into the control tower cab just as a King Air lifted off the runway. Above this King Air was an aluminum cloud in the form of a 747, gear down, flaps hanging and ready to squat like some prehistoric monster chasing an egg snatcher from her nest.

The King Air got the word and left. The controller who orchestrated the show received a lesson in the value of not running too tight a slot when an airplane the size of Belize is involved.

As much as possible, we try to find a route for the larger aircraft, set them to it and, hopefully, never change the sequence. If a smaller airplane comes into the picture, it often becomes more expedient to put the smaller aircraft in trail of the larger one and suffer the complaints of the inconvenienced pilot. But if you feel as though you've been truly wronged and should have been ahead of the airplane you followed, then call the tower. Demand answers!

You'll be connected with a GS-14 staff flak catcher trained in the art of sounding sincere on the telephone and promising to look into the matter while completely ignoring you. Seriously, they really won't ignore you. Most FAA facilities will take complaints seriously. They might not be able to do anything about it but they'll at least explain why it happened.

There is justice

As maligned as the GA pilot may feel at the hands of ATC, there is a bright side. If you're operating under Part 91, you have a distinct advantage over the air carriers; the right to depart or shoot approaches when weather is below minimums.

For two weeks last winter, Des Moines had been blanketed with low ceilings and even lower RVR, effectively halting air carrier operations. Like beached whales, the Boeings and DC-9s sat at their gates, pilots staring at the fog, thinking of new excuses to pass along to the passengers and occasionally watching the odd Bonanza or Cessna taxi through the murk on the way to the runway.

"Bonanza 123, runway 12L RVR zero, cleared for take-off."

"Mooney 123, runway 12L RVR zero, cleared for the approach."

I'm not suggesting you go around busting minimums. But one thing's for sure: The airlines live and die by the almighty RVR and more than once, the thing is several minutes (or more) behind reality. So, when you've received your approach clearance, and just prior to melting into the top of the overcast on the glideslope, look at all the Boeings circling in the hold. Smile to yourself and give them a wave.

There's a lot more in the controller's bag of tricks that he or she can use to expedite matters. One of the most useful tools for both pilot and controller is the pair of Mark I eyeballs that come as standard equipment on every pilot.

The pilot can use his or her eyes to make a flight faster and more direct. The controller can use the pilot's eyes to help solve some potentially sticky separation problems and even to avoid the dreaded "deal." The official name for this is visual separation, and it can be a positive boon to controllers. Tracon controller Paul Berge explains how in this section.

"Maintain Visual Separation...."

The non-flying masses are marginally tolerant of those of us who dare leave the earth. Two misconceptions have kept us free of further government meddling: a) The public thinks we're all on flight plans, and b) the FAA has "air controllers" at every airport directing every airplane everywhere through the magic of radar. Let us pray that no one outside aviation ever discovers that the reason windows were installed in cockpits was to see and avoid other airplanes without the use of TCAS, radar or other forms of electronic wizardry.

As airspace becomes more restricted and the cockpit evolves in complexity, there remain a few procedures in the system designed, as the AIM says "...to reduce pilot/controller workload and expedite traffic by shortening flight paths to the airport." Visual separation and visual and contact approaches are among them. These are basically IFR procedures that rely on a mixed bag of visual and instrument separation standards. Short of the pilot actually cancelling IFR, they allow controllers to cram more airplanes into a given space.

Do-it-yourself ATC

If every pilot requested visuals and contacts or even VFR climbs or descents, ATC could be run by an answering machine: "Thank you for calling Kansas City approach; all eastbound altitudes are busy; if you wish a westbound IFR altitude, please climb VFR and maintain four, six or eight thousand feet. Upon reaching altitude, press star, then pound and your altitude. Thank-you." We may be flying that way sooner than we think but in the meantime, traffic separation is a ducking and weaving game, failure at which brings more than heartburn.

Visual separation begins the instant you call ground control for taxi, even on days when the weather is so low that *nobody* should be flying. Larger airports may have ASDE ground radar but most controllers agree that it's marginal at best. The controller in the cab wants to see your airplane to clear you to a runway. If you're hidden by a hangar, your position report, "...at Van Dusen," serves until you can be spotted and told who to follow.

Once at the runway, the local controller (the tower) has the option of applying visual separation between you and any other arrivals or departures. When you're told to "taxi into position and hold," there may be no radar available, in which case the tower is doing what you would do at an uncontrolled airport; watching the arriving traffic and visually judging the required separation.

Weather permitting, the tower will clear you for takeoff and, in the same breath, tell you to, "...maintain visual separation with the departing twin Cessna." The controller judges that your aircraft will remain reasonably clear of the twin until radar separation can be established. The controller is relying upon the pilot to keep the twin in sight and not hit it. The instant the pilot acknowledges the clearance, the burden of separation responsibility shifts from tower to pilot. Well, mostly.

Behind the scenes

When the tower controller issues the clearance to "maintain visual separation" he has taken into consideration all factors which might preclude the pilot from being able to follow the instructions. Obviously, weather is a factor. The controller is supposed to consider what's being reported on the ATIS and what he or she sees out the window. Maybe there's a fog bank moving across the shoreline that may swallow the Cessna, causing the trailing Cheyenne to lose sight of it.

Controllers must consider aircraft performance, too. Perhaps the Cheyenne will outrun the Cessna before diverging courses can be established. I once cleared a Boeing 737 for takeoff and, with minimal separation, told a Lear 55 (I'd never seen one) to "maintain visual separation with the 737; cleared for takeoff." The 737 must have been heavy or had a door open, because it climbed like a sick Aztec. The Lear ran up its tailcone, until the pilot radioed: "Now, what, tower?"

Occasionally, a less-than-razor-sharp pilot may accept a clearance requiring visual separation by simply rogering the controller without realizing the consequences. An alert controller will sense trouble ahead and decide on his own to delay the second airplane's departure.

Ultimately, however, it's the pilot's responsibility, under FAR 91.113(a), to "see and avoid" other aircraft, weather permitting. Regardless of what punitive ramifications may befall the controller, the more serious consequences of not complying with a visual separation clearance are always reserved for the pilot, or his estate.

Ignorance is bliss

The tower may also apply visual separation without either pilot being aware of it. As long as the controller can see the two aircraft through the tower windows and swear on a pile of 7110.65s that there's blue sky between them, it's legal.

If, however, he or she loses sight of either aircraft, some other form of separation had better be provided, such as quickly shipping one aircraft to another frequency before the pilots get wise. This is referred to as "frequency separation" and is difficult to find in the manual.

An approach controller, vectoring himself into a corner, may call upstairs to the tower: "Hey, Rachel, can you see the Citation and the Cherokee?" A cooperative tower controller will then answer: "I can provide visual separation; put them on my frequency." Radar service ends and eyeball vectoring takes up in a desperate effort to salvage careers.

Still, visual separation is not a Get Out of Jail Free card for controllers. It's reserved primarily for terminal (and not Center) controllers when, to quote the .65, "other approved separation is assured before and after the application of visual separation."

This means I cannot vector your Mooney on an intercept heading with United's 727, at the same altitude, and then, with both airplanes exactly three miles apart, ask: "Mooney 2PF, do you see the 727?" I must have radar separation *guaranteed* before I try for visual separation. In a similar situation, I could vector the Mooney at an angle approaching the jet, then at some point, say ten miles, ask the Mooney pilot if he sees the jet. If the answer is yes, I say: "Maintain visual separation from that traffic."

If the two aircraft are on a converging course, I'm required to tell the pilots: "Traffic, one o'clock, eight miles, eastbound, a Boeing 727, five thousand, *on converging course*." After the Mooney agrees to avoid the jet and the courses are still converging on radar, I then advise the Boeing pilot, "traffic has you in sight and will maintain visual separation." I'm not certain what comfort it is to the 727 crew that a Mooney pilot promises not to ram them but hey, I'm just doing what the book says.

Keeping two aircraft separated by altitude until one pilot sees the other is usually the easiest route. A radar controller can put one target directly over another, ask the pilot to look down, and "...maintain visual separation with the 727 below, descend and maintain 3000." If they can't see each other, vertical separation is still assured.

Remote view

Visual separation authority can be delegated by a radar approach facility to a neighboring non-radar tower and there are lots of them around. An approach control facility may own the airspace overlying a VFR tower's ATA and control both the IFR approaches into the airport and the IFR releases.

Often, approach may not have radar coverage to the ground—as was the case some years ago when I worked at Monterey approach. Salinas tower would call for release of an IFR departure before an IFR inbound, already on tower frequency, had cancelled. Salinas, a VFR tower, would be authorized by approach to provide visual separation between the two so it could then release the IFR departure.

The release responsibility passes to the VFR tower, who clears the departure, guaranteeing that it will be clear of the IFR arrival prior to reaching departure control frequency. This rule is *never* applied to Flight Service. An FSS may call approach or Center on a landline for a clearance, which they repeat verbatim to the pilot. But release is always at the discretion of ATC.

Flight Service may tell approach, "Mooney 2PF just cancelled his IFR with us and Cessna 89B is looking for an outbound clearance." In the old days, when FSS had windows and specialists wore skinny black ties and could actually see the arrival land and ask for the release. That doesn't happen much anymore. There are few FSSes left, and they're usually relegated to the outskirts of the airport anyway.

Use your eyes, and your head

There's more to this business of using your eyes while IFR, of course, and it can be just as useful to the pilot as to the controller, considerably expediting his or her arrival at the destination. I'll leave that discussion to the section on getting favorable routings, however.

One of the things to remember about visual separation is that it makes our lives as controllers a lot easier. When you hear us ask if you see some traffic, remember that if you can spot the other guy and keep him in sight you'll probably be doing the controller a real favor.

Holding is a procedure that gets a lot of attention during instrument pilot training, but in the real world it doesn't actually happen all that often. One of our editors, in nearly 300 hours of instrument flying, only had to hold twice...and one of those was to get out of the way of Air Force One.

Still, holds do occur. Controller Paul Berge tells us why they happen, and how you might be able to avoid one.

Handling Holding

Ever made dinner reservations for 7:00 p.m., arrived at the restaurant at 6:56 and, upon giving your name to the maitre d', discovered that not only is your table unavailable, but you can't even get near the dining room? Failing to grease the upturned palm with a ten, you find yourself in the lounge—waiting. In restaurant terms, you've been slapped with a hold.

It's the same with Air Traffic Control. You see, prior to becoming controllers, most of us worked our way through the FAA's Academy in Oklahoma City waiting tables in the school's cafeteria. There, we were taught not only to keep a dozen pilot requests straight, but we learned the importance of fielding complaints.

ATC, unlike the restaurant business, serves its patrons on a first-come, first-served basis, or so you've heard. Deprived of the chance to slip a ten to the controller ("Psst, let's say we skip the hold and go direct, eh?") you're forced to wait your turn. In a busy environment, someone will be first, and everyone else will follow. When the conga line gets too long, someone has to wait in the lounge.

Half our purpose in ATC is to get airplanes from the sky and onto the ground (preferably at their destination airports) as safely and quickly as possible. Holding interrupts the flow, but it's sometimes the only alternative.

The reasons for holding are as varied as the number of controllers in the system. Holding is not the norm where I work. I can't speak for busier facilities, but if I initiate holding, my troubles are just beginning.

Holds definitely have a regional flavor. A coastal airport, for instance, might be routinely fog-bound until mid-morning, clear all afternoon and W0X0F again at sunset. To avoid holds, local pilots schedule themselves accordingly.

The midwest has a less predictable day. A cool morning with gentle southerly breezes can lead to a wicked afternoon with towering cumulus building into thunderstorms. Winds shift with each passage. The runways disappear under heavy rain showers, and pilots may find themselves unable to even approach the field.

For controllers, the problem of where to put airplanes is compounded by the weather. As thunderstorms grow, airspace vanishes. A handful of airplanes are squeezed into tighter places while voices on both ends of the frequency grow strained.

"Cleared to the (fix); hold as published..." should be the easiest solution. But what if the published hold sits in the middle of a level 5 cell? A new holding pattern must be created. All bets are off; it's the controller's chance to be creative. A clearance direct to the VOR might be easiest, or, my favorite, a DME fix on the inbound radial you happen to be on. I've even asked pilots to hold over ground reference points, such as lakes.

Just because you're on an IFR flight plan, doesn't mean we can't negotiate a simpler way to park you. I've asked pilots to simply make a couple of 360s to buy time. Of course, there's no EFC but in the event of lost comm, I'd expect them to follow normal lost comm procedures, via last route cleared, or expected, and so on.

A winter snowstorm can clog the system for hours. Once the visibility lifts, the runways are usually packed with snow. Estimates of runway opening time are always optimistic. The result: Everyone holds, maybe on the ground at the departure airport or, less desirably, airborne. In this scenario, the easiest way to hold is to stick everyone on the localizer, at separate altitudes and, as runway conditions allow, peel off the arrivals one at a time, stepping everyone down in the process.

How a pilot enters or flies the hold is almost moot because everyone is separated by altitude. In more complex airspace where holding might be in progress at a number of fixes, the pilot's ability to remain within protected airspace could become a more serious consideration. A target wandering away from the hold could draw a gentle, "Where are you going?," followed by a hand-holding vector back to the corral.

Can I tell if your hold entry is FAA/AIM/Martha King-approved? Teardrop, parallel or direct? Maybe. Do I care? No. As long as your radar target is mushing around the vicinity of the fix and at the altitude assigned, I don't get too concerned.

If you actually find yourself compelled to hold and you haven't the foggiest notion how to enter the pattern, don't panic. There's no air traffic snitch in the radar room measuring your turns or timing your inbound legs. Just jump right in, hold altitude and thrash your way through it the best you can. If you get bored with right turns, ask for left. If you want longer legs, speak up.

Listen to the ATIS to glean what you can regarding delays. If holding looms ahead, ask for a speed reduction. Maybe you'll never get to the hold and even if you do, it can't last forever. Remember, in more than 50 years of air traffic control, we haven't left one up there yet.

As Paul Berge just noted, one of the reasons holds occur is weather. In the bad old days, weather caused more problems than it does now, simply because now the system tries to compensate for the effect of bad weather at the destination through flow control, gate holds, and so forth.

One of the programs put into place to keep severe weather from playing havoc with ATC is SWAP (Severe Weather Avoidance Plan). It does help, at least for a while, and at least in major terminal areas. But, like the other congestion remedies instituted by ATC to keep things from getting too badly piled up, it can get overloaded all too easily.

Des Moines controller Paul Berge tells us what it's like in the "boonies," where these programs don't help much, then goes on to explain SWAP and its limits even in major areas.

How We Deal With Weather

It's a recurring nightmare. First, all the airplanes in the world show up on my radar scope. Then, just as I'm formulating a plan—a way to get them all in line—the scope begins to shrink. Only this is no dream. It's spring in the Midwest, when some of the world's most entertaining thunderstorms arrive.

Predictably, they first appear as tiny light blobs on the southwest corner of the scope, like innocuous spots on the X-ray of a smoker's lungs. Nothing to worry about; they'll go away. But they don't. They grow and eat the scope, until there's no room left for airplanes.

As the front nears, the spots become globs then blossom into squall lines stretching ever wider, drawing a Great Wall between Omaha and Des Moines. National Weather Service spits out convective sigmets, giving a play-by-play of the coming event. Center is on the phone explaining why United from Denver has deviated over Perth Amboy, New Jersey, and the pilot wants to know how our weather's holding up.

We lie. We say it's great inside the tracon; smooth ride.

It's gets worse, always

Convective weather has a habit of stopping periodically to gain strength, as though drawing energy from the countryside by flipping over trailer courts. On radar, it looks like an army is deploying troops for an assault. Their bivouac area covers four counties—midwestern counties, each the size of Connecticut.

Eventually, the gods figure they've gathered enough strength, and the line recommences its inexorable march through our airspace. Another sigmet—"24 Central"—hits the printer, warning of wind gusts to 65 knots and hail the size of prairie oysters. Remembering a passage from an Ernest Gann novel, I sink lower in my controller chair and flick a cigarette lighter on and off in front of my face. I'm not sure what that's supposed to do, but it worked for Ernie; he sold millions.

With the weather still west of the field, the Chicago arrivals slip in from the east without trouble. On final, they comment that there appears to be the mother of all thunderstorms brewing to the west. About that time, a Seneca driver checks on the frequency, just west of the line, asking if we can see a break in it. I explain that our radar (ASR 4, only recently upgraded from steam to diesel) isn't of much value distinguishing good weather from bad, and, frankly, it all looks bad out west.

He agrees and presses on. A few air carriers sneak in from southwest, picking their way through the peaks, reporting "...moderate chop...Jeez! Make that severe!"

Where'd he go?

The first crackle of static from lightning is in my headset, and I watch the Seneca disappear into the backside of the front. His target simply vanishes in the greenish white smear where the storm is. We switch the radar from linear to circular polarization in an effort to cut out some of the clutter. On CP, we can see the beacon targets again—at least the ones outside the heaviest precipitation—but we sacrifice what little ability we had to see weather.

We also lose target strength, relying more heavily on the secondary beacon return. Which means, among other things, we won't see non-transponder targets too well. Of course, non-transponder pilots are probably smart enough to keep the Champ inside the hangar. It's a toss-up. We'd rather see airplanes; the pilots can tell us about weather. Circular polarization, however, doesn't filter out the heaviest precipitation, so if we're operating on CP and can still see weather on the scope, chances are it's mean.

Still, the airspace contracts as pilots announce they have no intention of taking our headings. Like ants on a sinking log, they all find themselves heading for the one remaining dry spot in hopes it leads to safety. It's then, that tower reports: "Visibility two miles; heavy rain and lightning off to the west."

We pass this to the pilots, who respond with the enthusiasm of the doomed. Inside the tracon, it's still smooth and dry, but more and more airplanes are trying to occupy a continually shrinking airspace. Each plane that makes it through the line sends back a report. If a smooth ride is mentioned, everyone wants to know where the hole is, and will it be there when they arrive.

Rarely, it seems, does anyone say, "I think I'll turn around and sit this one out." Freight haulers, in particular, never turn tail. Mooney pilots opt to punch into thunderstorms, reverse course midway and punch their way out to try again.

No visuals today

The first thing to go is the visual approach; everyone lines up for the ILS, and everyone, understandably, wants to be first. On a normal day, a controller can vector the slow guys to make slots for the jets, but when thunderstorms are etching away at the localizer, no one takes our headings.

Speed control becomes the only tool when the kerosene arrivals have to follow the Cessnas, and the last thing a pilot wants to hear when outrunning a squall line is, "...reduce speed."

Not only does airspace evaporate as the front passes, but—in textbook fashion—the winds shift, so that everyone, carefully lined up for the ILS to 31R, suddenly hears about winds out of the southeast at 20 knots, gusts to 40. It's time to move the party into another room. Pilots can figure an additional 25 flying miles to be vectored to the reciprocal ILS, and hope when they get there the winds don't shift again. I once taxied a YS-11 to four different runways, until the pilot threw up his hands and said, "...just clear me for take-off; I don't care what the winds are! I'm running outta fuel!"

Forget the book

Creativity becomes the watch word as airspace is swallowed by weather. Where the book might require a controller to turn the aircraft onto the ILS no closer than two miles outside the approach gate, a healthy thunderstorm sitting over the gate mandates a different plan. It's not unusual for pilots to ask to intercept at the marker or even inside the FAF to avoid weather.

A controller willing to stick his close-to-retirement neck out may vector a pilot onto an improvised surveillance approach just to keep the pilot out of what can be seen on the scope. The danger here, of course, is that should the controller misjudge and slam the aircraft into an unseen build-up, the courts may not be terribly sympathetic.

Drastic times require drastic innovation, and the legalities of such an approach can be hashed out later when everyone's on the ground and, hopefully, not under it. Essentially, what the pilots are executing is an abbreviated makeshift approach without any intention of a missed. The niceties of "legal" technique become moot when a pilot is pleading for an escape route.

Larger facilities may rely upon SWAP or other acronyms to ward off disaster, but here in the outback of ATC, we simply take what Center sends us. Sometimes—without warning—they hand us what Chicago can't swallow. If you want to see something funny, watch a planeload of passengers, originally bound for O'Hare, trying to fathom why-in-hell they're in Des Moines. Not only thunderstorms, but snow or fog can divert big city traffic to alternates in the middle of nowhere. There's no "plan" implemented; they simply appear. The creates havoc with the cattle we normally let graze along the taxiways between flights.

Where we may see SWAP-like plans implemented is with our departures headed for DFW, ORD, DEN or MSP. When the big houses get swamped, the backed-up theories go into effect, and places like Des Moines are told to ground stop all departures. We usually get these flow messages just as American and United are both taxiing for the runway. It's then up to ground to explain to the irate pilots, "...well, they don't tell us why; Center just says you can't go...no, I don't know when you'll get out...no, I don't know why, either. I'm only a GS-12, Captain!"

When releases do trickle in from Flow, they may include reroutes. A Chicago-bound flight which may have filed over Dubuque (DBQ) may be handed a southerly routing over Bradford (BDF). We usually give the pilot the reroute, then say, "And Center needs you off in 60 seconds; cleared for immediate take-off, you can figure out the reroute later."

The success of our entire ATC system is based on two premises: 1) Controllers won't go on strike (again), and 2) It's a big sky out there, so it's really hard to screw up. Rule number two, the Big Sky Theory, applies to VFR and non-convective IFR days. In facilities such as Des Moines, we don't warrant the status of telling Flow Control to shut off our arrivals or run everyone over Wyoming before they get to us.

Smaller ATC facilities are like fire stations. We sit around reading comic books and watching Murphy Brown until someone tosses a match into the excelsior, then it's madness until the flames are out. When thunderstorms hit Des Moines, it's time to toss out the 7110.65 and squeeze ten pounds of aviation into a five pound scope.

How it's supposed to work

Every major terminal has its own scheme for moving traffic but all have one thing in common: they herd airplanes through fixed arrival and departure paths. When thunderstorms move into those paths, it's like pinching a garden hose. The water will flow for a little while, but eventually, there's just no place for it to go.

ATC is pretty good at rolling with the punches when weather rolls in but the fact is, at many terminals, there are just too many airplanes and too little airspace. During a major departure push, the system is already peaked out. Even a harmless looking level 4 cell can push it into momentary overload. A persistent line of thunderstorms will bring the whole show to a crashing halt and it may take many hours to put it back together.

When the weather is really nasty, ATC fights a rear guard action to keep any traffic moving at all. Usually, the first maneuver in what may be a losing battle is SWAP, for Severe Weather Avoidance Plan. SWAP is supposed to be available everywhere but it's mainly used around major terminals and their satellites, and then only in regions where thunderstorms are common.

Theoretically, SWAP is transparent to pilots. Other than a brief mention in the glossary, the AIM has nothing to say about it. Occasionally, though, you'll hear SWAP mentioned on an ATIS broadcast, along with specific instructions and requests for pilots.

What it does

Like ATC's other flow control remedies, SWAP is kind of a proactive stop-gap measure. It's supposed to minimize the minor chaos caused by a single cell camped on a departure fix and stave off total collapse when a line of storms sweeps across the departure corridors. When it's in effect, SWAP gives controllers the option of revising the usual preferred departure routes to avoid the weather.

On a typical summer day, ATC starts considering SWAP long before the weather moves in. Early in the day, the TMU controllers in every facility in the country likely to be affected by severe weather, confer by teleconference. They review the day's forecast weather and traffic levels and discuss various options should the ugly weather actually materialize.

In years past, these options—usually reroutes—have been ad libbed but lately, SWAP procedures have become more formal in some Centers, to the extent that they're actually computer-stored preferred routes. In the northeast, for example, dedicated preferred routes (unpublished) exist strictly for use when SWAP is in effect. Besides reroutes, ATC has the option of invoking its everyday flow control tricks; namely ground delay programs (gate and ground holds) and in-trail restrictions.

Timing is the critical factor. In the midwest and northeast, thunderstorms seem to develop in late afternoon, a period that corresponds with departure pushes. Planned ground delays knock the sharp edge off the peaks which, theoretically, opens up more airspace so SWAP can keep things moving.

Going to SWAP

As a rule, Centers and tracons don't like going to SWAP and they'll avoid as long as they can. Even though it's supposed to prevent delays, SWAP is a departure from the routine and it necessarily slows things down. When small areas of weather block only one fix or route, controllers may have enough airspace to allow aircraft to deviate around the cells, without resorting to the massive reroutes that SWAP entails.

The "going to SWAP" message comes via telephone and teletype from the appropriate Center's TMU, whence it filters down to the radar room through the area managers and supes. SWAP works a little differently everywhere but basically, it puts all departures headed for fixes blocked by weather on "APREQ" or approval request.

That means the local tower or clearance position has to call Center for a release on each individual aircraft headed in a particular direction. At Kennedy, for example, the normal procedure is to taxi a platoon of aircraft to the runway while clearance works out a routing with the Center. If the usual preferred route is weathered out, Center will issue a last minute route amendment that may change several times before the aircraft is released.

SWAP is selective in implementation, meaning that reroutes usually apply only to aircraft heading for certain fixes. But that doesn't mean you're home free if the weather is west and you're going east. In the New York area, the major west departure fixes are ELIOT, PARKE, LANNA and BIGGY, which are arranged in a north-south line that runs about 60 miles west of Kennedy. The northernmost fix, ELIOT, is only 25 miles from the southernmost fix (BIGGY) so it's not unusual to have two or even all of the fixes blocked by weather. In that case, departures may be rerouted to the north fixes (GAYEL, HAAYS, NEION, COATE) and mixed in with the traffic that normally uses those fixes. It's the classic case of five pounds of dung in a three pound bag; everyone takes a delay.

In other parts of the country, there's more airspace and less traffic so even though frightful lines of weather come marching through during the afternoon, there's enough room to reroute without formally going into SWAP.

Meanwhile, down low

GA aircraft, even low and slow ones, sometimes use the same fixes as air carriers but often, they sneak around at low altitudes on tower-en route-control routes and/or they depart from satellite airports.

GA pilots do have certain advantages when SWAP goes into effect.

For one thing, satellites aren't clogged with air carrier jets looking for a place to park. Second, thunderstorms have a way of ridding the low-altitude structure of the faint of heart, meaning more routes are open. Controllers also know that GA aircraft will sometimes venture where air carriers will not. You may think that's due to ignorance or false bravado but actually, some GA airplanes are better at detecting storm threats than airliners are and air carrier pilots try to avoid even a hint of turbulence, lest the paying passengers knock their noggins on the overheads.

Probably the best way to avoid getting hung up by SWAP is to avoid it in the first place. If thunderstorms loom, check the current radar and see where the storms are relative to the local departure fixes. Also, log onto to DUAT and check the flow control messages. If SWAP is in effect, it will be noted there, as will any en route or destination delays.

We've all gotten vectors; in fact, that's the way the vast majority of real instrument approaches are flown. We know that the idea is to get you lined up with the final approach course at an altitude that will let you shoot the approach, all in order to avoid the procedure turn (and holding up everyone behind you while you execute it).

Usually, vectors work out just fine, with no complications. Every once in a while, though, something happens to interrupt the flow: perhaps the winds are stronger than expected, or the controller gets distracted and lets you blow through the localizer.

Knowing the whys and wherefores of vectors can allow you to avoid trouble when being steered by ATC. U.S. Air Force F-16 pilot J. Ross Russo explains.

Getting Vectored

If you're like most instrument pilots, you spent a lot of training time flying full approaches, muddling through procedure turns while you tried to figure out just exactly where the final approach course really was. Unfortunately, all that training wasn't good practice for the real world; the full approach just isn't the usual way of doin' business. Vectors to final is the way it's really done.

There's no doubt that full approaches are more difficult than simply following a controller's headings. But just because vectoring relieves you of getting yourself onto final, doesn't mean there aren't a few things you ought to be doing while the controller lines you up. There are some subtleties to being vectored and they change with the weather or a pilot's requests. Knowing a bit about the finer points can save you time and result in a cleaner approach.

First, let's look at the approach architecture from the controller's point of view, since he or she is doing most of the headwork on a vectored approach. A controller must consider a point on the approach that you won't find published on either Jeppesen or NOS plates. It's called the "approach gate" and it's defined in the pilot/controller glossary of the AIM as "an imaginary point used within ATC as a basis for vectoring aircraft to the final approach course."

On precision approaches, approach gates are generally established along the final approach course one mile from the outer marker (or another fix, in lieu of a marker) on the side away from the airport. For non-precision approaches, they're a mile outside the FAF. In either case, when measured along the final approach course, the gate will be no closer than five miles from the landing threshold. In some cases, they may be farther from the threshold.

Just how your vector relates to the gate depends on the weather. If the weather's good, the controller can vector you directly to the gate, but not inside it. For vectoring purposes, the FAA defines "good weather" as a reported ceiling and visibility of at least 500 feet above the minimum vectoring altitude (MVA) in the area you happen to be in plus a visibility of three statute miles. If no weather reporting is available, Pireps on ceiling and vis will do.

A "good-weather" vector to final should produce a *groundtrack* that will intercept the final approach course at an angle not greater than 20 degrees. Note the emphasis on groundtrack. To compensate for wind, the controller might assign a heading that will *appear* to result in an intercept of more (or less) than 20 degrees.

When the weather is less less than 500 feet above the MVA with less than three statute miles of vis, expect a vector to a point no closer than two miles outside the approach gate. In this case, the controller should provide a vector that will result in an intercept groundtrack that's no greater than 30 degrees. If it all works out, this should put you on final at least seven miles from the landing threshold.

Just ask

Even though controllers try to correct their vectors for winds, it's been my experience that they don't always succeed. Remember, the controller is looking at a radar-enhanced depiction of your aircraft, not a real-time representation of where you are. This lag, and the fact that you're not the controller's only customer, sometimes results in a bad vector.

One case in particular comes to mind. I was flying from Tampa to Jacksonville's Craig airport (CRG) one rainy, windy night. The approach controller was vectoring me to final for the ILS 32 approach. I'd been battling 60-knot winds out of the northeast for the entire trip, so I knew that the wind correction angle on final would be truly impressive, perhaps in the neighborhood of 30 degrees.

I was on a heading of 050 degrees, 90 degrees off the final approach course. This heading, along with VOR, DME, and LORAN readouts, confirmed that I was on a base leg southwest of the airport and that I had a direct headwind. The mental warning flags were already popping up. I knew this monster headwind would quickly turn into a monster crosswind on final. All the ingredients were present for an undershooting vector to final. All it would take was a slight oversight on the controller's part.

Since the weather was lousy, I knew that I'd be vectored to a point two miles outside the gate and that the stiff winds would mean I'd need a heading of about 020 degrees to get the required groundtrack for a 30 degree intercept. Would the controller be sharp and give me a heading that would correct for the wind?

"N1161X, three miles from ADERR, turn left heading 350, cleared for the ILS runway 32 approach."

I knew right way that this heading would never intercept final; it would only parallel it. So, acting as the final authority, I let the controller know that because of the wind, I'd need 020 degrees to intercept. The heading was approved as requested and the approach was completed.

Say the magic words

Okay, so we know that if the weather is a certain ceiling relative to the MVA, we'll get a longer trip down final. But how do we to tell what the MVA is? Unfortunately, only the controller knows for sure since he has above his radar screen a chart depicting all the MVAs in his area. Absent the chart, use the procedure turn or glideslope intercept altitude as a guide to the MVA. It should be relatively close. If you're really hard-over on the MVA issue, contact the plans and procedures specialist at your local ATC facility and request a copy of that facility's MVA chart.

About the only sure fire way to know where you're being vectored is to be cognizant of the visibility. If it's reported as less than three miles, you'll be going the long way around, two miles outside the gate.

There are some some things that we pilots can do to speed up or slow down the vectoring process. But as with other aspects of the ATC-pilot relationship, we have to ask. Controllers can't offer.

Let's say you're flying into your home drome and the visibility is reported as two-and-a-half miles. Bingo. Plan on a seven mile final. But you've flown the approach hundreds of times before, and you're familiar with the local terrain and obstructions. To shorten up the final, simply tell the controller that you'll accept vectors inside the approach gate.

This allows the controller to vector you to a point not closer than the final approach fix, with an intercept angle of not greater than 20 degrees. Depending upon your direction of flight, this trick could easily save you seven or eight miles of vectoring. Of course, your request has to fit in with the controller's overall sequencing plan. He won't be able to give you close-in vectors if they'll violate minimum separation standards with another aircraft.

On the other hand, even if the weather is good but you've dropped your timer, misplaced the plate or are otherwise trailing on the static wicks, you'll want to slow things down. Just tell the controller that the approach will be "coupled" or "evaluated," or words to that effect. You'll then get vectored to a point at least two miles outside the approach gate.

It may take a couple of requests to get the message across, however. As an F-16 flight examiner for the Air Force, I frequently fly evaluated approaches. Earlier this year, I was giving an instrument checkride to a pilot in Great Falls, Montana. The weather was good and the vectors were resulting in tight turns to final.

I wanted to see an approach that would approximate actual instrument conditions, so I had the examinee inform the controller that all subsequent approaches would be "evaluated approaches." Approach rogered our call but the next vector was as tight as the first.

Obviously, we weren't communicating. During climbout I informed the controller that the next approach would be an "evaluated, coupled approach," and I'd like vectors two miles outside the gate. It didn't matter that the F-16 can't fly a coupled approach. All that mattered was that I used the magic words. All of them. It worked.

Heed the clue bird

When it comes to the radio, pilots are lucky; we can use just about any phraseology we want. Controllers, technically speaking, aren't as fortunate. Their words are very strictly prescribed by the Air Traffic Control Manual (7110.65). And every so often, an FAA boss listens in to evaluate how a controller is speaking.

If a controller, for example, sees that an aircraft is well off course while inside the approach gate, he has no option other than to say: "N12345, X miles from the airport, X miles right/left of course, say intentions." These words are spelled out for him, crystal clear.

They should be just as clear to you should you ever hear them while on final approach. The clue bird is hammering desperately on your windscreen. The controller is saying everything within his legal limits to let you know that this particular approach is not going well for you. When the controller asks for your intentions, request vectors around for another approach. When things have gone that far awry, they seldom get any better. Take the hint. This is one birdstrike you'll be thankful for.

Vectors across final

What should you do if the controller forgets about you and drives you across the final approach course without clearing you for the approach? The controller's manual is very clear on this point, too. The controller is required to inform a pilot if a vector will take him across final, along with the reason. Something like this; "N12345, expect vectors across final for spacing." But, if for some reason the controller is unable to inform the pilot, the pilot is not expected to turn inbound on the final approach course.

While receiving vectors, the headings that you're given, in conjunction with what your avionics tell you, will give you the position awareness you need to anticipate the controller's next action. If that 30 degree intercept heading isn't accompanied by the approach clearance or a reason to expect vectors across final, it's time to launch your own clue bird. Speak up early and often. Ask the controller if you're to expect vectors across final or if you're cleared for the approach. Make your communication clear, timely, and definite. Clue birds fly equally well in both directions.

• Section Four •

Communications

Communications the "Right" Way

Possibly the most important skill an IFR pilot can have is that of good communications. But what, exactly, is "good?" As with most things, there are a great many ways to do it "right."

Following the AIM to the letter is certainly one good way. After all, the communication guidelines found there were written for good reason, and have withstood the test of time. It all still works, and works well. But that does not mean it's the most efficient way to talk on the radio in all instances. On any flight you're likely to hear a great deal of non-standard phraseology used. On occasion, it's even somewhat more efficient than the "traditional" phraseology you'll find in the AIM.

In this chapter we present two widely differing views of just what makes for "good radio." First we'll hear from J. Ross Russo, a U.S. Air Force fighter pilot with a fundamentalist view of what constitutes proper communications.

The Conservative Communicator

Let's get something straight from the start. When it comes to radio communications, I sit somewhere to the right of Torquemada. I freely admit to being a radio zealot and I'm proud of it. Although I stop short of going door to door on Sunday mornings, color me the phraseological equivalent of an Orthodox Jew, Southern Baptist and an old-time Catholic who still feels the Mass should be said in Latin. We all have access to "the word" as contained in the Airman's Information Manual. We've got no business trying to change it. Period, dot, end of story.

We've all heard radio Bozos spooging the frequencies with needless, nonstandard chatter. They're certainly being innovative and creative, but at what cost? Communication, clear and concise, is absolutely, undeniably fundamental to the effective and safe movement of air traffic. There are few absolutes in aviation, but this is one of them.

To the extent that communication capabilities are limited, the entire system suffers. Clutter the frequency with colorful colloquialisms and superfluous drivel and you're guilty of an abominable crime against all of us who fly. If you think it's cute, it's not. But there is a word to describe it: unprofessional. It labels you as a rube.

The biggest offenders? The airlines and the military. From what I've seen, the military jocks are even worse than the airline pogues. The reason you don't hear how bad the jocks are is because most military aircraft have UHF radios. That means you can hear what controllers are saying to the military traffic but you can't hear their reply. Good thing.

While military training does a good job of emphasizing radio discipline, precious little attention is paid to the AIM. I'd wager that the majority of pilots flying around in football helmets have never even heard of the thing. Is it any wonder, then, that these phraseological heathens are sinners? How can they be expected to know The Way if they've never seen The Word?

In every form of human endeavor, standards exist to measure performance. From the Olympics to dog shows, we have accepted rules and procedures. Aviation is no different. Our standards for acceptable communication are embodied in the AIM. But how many of us actually read the AIM, let alone subscribe to it?

We tend to imitate the things we hear the big boys say, or worse, strike out on our own. I cringe every time I hear someone use slang phonetics such as "Three Mini Mouse," or "Three Four Charlie Brown." My hat's off to the dutiful and professional controllers who consistently answer those calls with the proper phonetics.

I'm sure there are those who'd say, "Hey, what difference does it make?" Not much, unless you're someone who strives for artistic expression of his craft. Then it's a big deal. For example, let's say you own a red Mercedes, and you'd like to have it painted yellow. You could slop some hardware-store paint on the thing but would you be proud of it? Probably not.

The point is, it isn't difficult to make good, solid, professional, "religious" radio calls. It's simply a matter of discipline. Chapter Four of the AIM deals with Air Traffic Control, and Section Two of that chapter deals specifically with radio communication's phraseology and techniques. If it's been a while since you've been "to church," take some time to get reacquainted. The AIM is a fairly easy document to obtain: you can even order a personal copy from Sporty's.

But for those of you who would prefer the condensed version, allow me to offer the five most salient points. This religion is more condensed than most owing to the shorter attention span of its congregation:

1. It's essential to acknowledge each radio communication with ATC by using the appropriate aircraft call sign.

2. On initial contact, transmit the name of the facility being called, followed by your full aircraft identification, and then the type of message to follow (or your request if it's short). Call signs should never be abbreviated on initial contact (or at any time when other aircraft call signs have similar or identical sounds, letters, or numbers).

3. Use the proper format when saying letters and numbers. "Out of two point four for four point four" may sound like the heavies, but it's anything but righteous.

4. Precede all readbacks and acknowledgments with the aircraft identification. This aids controllers in determining that the correct aircraft received the clearance or instruction. The requirement to include aircraft identification in all readbacks and acknowledgments becomes even more important as frequency congestion increases.

5. Since concise phraseology may not always be adequate, use whatever words are necessary to get your message across.

You'll note that the fifth commandment, er, point, seems to neutralize the preceding four. Not so. If you're conducting business as usual and standard phraseology applies, then use it. If, however, the situation is unusual, then you're morally correct to wing it. It says so in the AIM.

In conclusion, just let me say that the FAA didn't pull their communication techniques and procedures out of thin air. A great deal of research went into discovering exactly how to get messages across with the highest probability of understanding. The entire system depends on that unerring understanding.

We've merely scratched the surface of the guidance offered by the AIM. If flying is your passion, your art, your craft, then by all means endeavor to perfect it, especially in the eyes of those who judge you only by the quality of your transmissions.

Denny Cunningham is a controller at O'Hare who has a different view of communications, one based more on individual judgment and flexibility than a strict adherence to "the Book."

Go With the Flow

How many times have you heard it said that flying, any and all of it, is about judgment. You're expected to exercise judgment about everything from airworthiness to whether Mother Nature has cooked up a weather system that's too powerful to challenge.

Considering the importance of judgment, I find it puzzling that some pilots abdicate their judgment to bureaucratic whim when deciding how to use the radio. After all, choosing the right words to get the point across is something we've all been practicing since childhood. I'd like to think most of us have a firm grasp of the concept by now.

I'll admit that not everyone exercises good judgment in communications. When I hear a pilot say "Hello down there at Chi-town Center, this here's good old Skybus 859, we're level at nine and feelin' fine," I want to crawl through his radio, reach into his throat, and pull his tonsils out by the roots. But, this is the kind of guy who would break wind and scratch his butt at a White House reception. He's going to be incorrigible in any situation.

However, there's a big difference between talking like a trucker and justifiable deviations from book phraseology. I agree the tried and true does just fine for some operations but much of the rest of the phraseology in the "book," (meaning the AIM and 7110.65, the controller's handbook), is fair game. Some of it's dated and unnecessary and some of it just sounds stilted. Often, the book just doesn't contain the words to pass the message we want to convey. Even the nimrods who wrote the book concede the point.

AIM, Paragraph 4-31b. reads, in part: "...the controller must know what you want to do before he can carry out his control duties. And you, the pilot, must know exactly what he wants you to do. Since concise phraseology may not always be adequate, use whatever words are necessary to get your message across."

Here, here. Now, wouldn't it have been nice if they had kept that paragraph in mind when they wrote the rest of the silly thing? But they didn't.

How else can you explain the Phonetic Alphabet? It's clear that this stuff was written by a refugee from World War II, an era when throat mikes made pilots sound like Donald Duck. Given the clarity of modern radios, the best way to get across the idea of "three" or "five" is to say "three" or "five." Follow the book, though, and you'll say "tree" or "fife". And try, just try, to keep a straight face when you use the approved pronunciations for "Oscar" and Victor" on the radio— "Oss-Cah" and "Vik-Tah" sound more like dialogue from an old Bella Lugosi movie than pilot-to-ATC communications.

Sometimes the desire to ad lib comes from a personal conflict with the way the muckety-mucks have deemed to officialize something. For instance, paragraph 4-12d says to inform the controller of the ATIS broadcast received by using the phraseology "Information Sierra received."

Say what? Who talks like that? What's wrong with "Have Sierra" or "with Sierra"? Obviously, not much, because two pages later, in paragraph 4-16b5, they use the words "...with Information Bravo" in another example. Apparently, they didn't think much of their original advice, either.

Often, you need to vary from book phraseology to say something different from what the pilot or controller expects to hear. In normal ATC communications, the pilot often knows what ATC is going to say next. The controller, in turn, expects a standard response from the pilot under most conditions.

But what about those situations where one of the parties is going to deviate, however so slightly, from the script? Isn't it reasonable to alert the other party that "Hey, fella, what I'm about to say ain't what you think it is, so you'd better be paying attention?"

Examples abound. Here's one from my own recent experience: Ground frequency at ORD has been 121.9 for years. Without even thinking, pilots switch to it upon clearing the runway. Controllers are used to it, too. The litany of "Turn-left-at-the-high-speed, contact-ground-point-nine" is so automatic, they don't even realize they're saying it.

So what happens when the tower's 121.9 receiver goes belly up in the middle of a rush? I'll tell you what happens; the substitute frequency (in this case, 119.25) may as well be located in Yugoslavia for all the use it gets.

"Turn left at the high speed, contact ground 119.25," says the controller. But that's not what the pilot hears. He hears what he expected to hear, which is "ground-point-nine." So he clears the runway, and starts transmitting into the empty void where 121.9 used to be. When he gets no response, he tries the other radio without success, then decides maybe he's got a problem with his headset. After five minutes or so he gives up and goes back to the tower to tell them that he can't get an answer on 121.9.

Chances are he won't get a response from the tower, either, because the local controller is too busy dealing with the go-arounds resulting from a blocked taxiway, not to mention the shouting match taking place between the ground and local controllers as they try to figure out who's talking to the Bozo blocking the taxiway and why ain't he moved yet.

The better way is to make that "routine" transmission not routine. Here are a few successful examples: "...new ground frequency is 119.25." Or "...ground frequency is broke, call 'em on 119.25" or even

"...contact ground on the strange and unusual frequency of 119.25." Most pilots chuckle and get the new freq the first time. It's not approved but it works.

How about those times that the controller or pilot needs a little something extra from the other guy? On a tight "gap shot" in a crossing runway operation, there's often a window of only 15 seconds when a departing aircraft can start takeoff roll without T-boning traffic landing on an intersecting runway. A routine "Taxi into position and hold," followed by a laconic "Cleared for immediate takeoff," doesn't convey the immediacy. A better method is to "give the pilot the flick" and thereby assure that the flick doesn't turn out to be a disaster movie.

"Gargantuan 453, position and hold on 14L, be up on the power and ready to roll immediately behind the landing traffic on 22R; there's traffic two miles behind him." Most pilots will deduce that if they don't move their tail they're gonna end up as a hood ornament on somebody's airplane, so they're on the roll just as soon as they hear "cleared for takeoff." This isn't in the book and it makes the bureaucrats crazy but it works.

As in all good things, however, discretion is the key. Some ad lib phraseologies convey meanings that aren't intended, such as the time a controller told an air carrier departing runway 4L at O'Hare to "turn right to ninety." The book says (with good reason) that he should have said "turn right heading zero-nine-zero".

By the time the 727 had completed a 250-degree right turn and steadied on a final heading of 290, he had climbed through the final approach courses of three arrival runways and became "Aircraft Number 1" in four operational error investigations.

In this case, there was no reason to depart from the standard. This is simply a case of an individual exercising poor judgment. But should such a case be cause for a blanket indictment of all non-book phraseology? Of course not. The point is to educate pilots to use good judgment, not eliminate the concept of non-standard procedures.

I say be professional, sure, but also be creative and innovative. Use your head and temper your talk with good judgment; say what you mean and leave no possibility for misunderstanding. As long as those bases are covered, the phraseology back door is wide open, and we should be running through it full tilt.

So, who's right? Both of these contributors are skilled professionals who deal with the air traffic control system every day. Both are correct: The AIM is the right way and will always get you through, and it also can be inefficient and awkward.

The bottom line when it comes to IFR communications is that if you're not sure, use the AIM guidelines. You won't go wrong, and you can't get busted for following the book. If your radio technique is sharp enough to deviate in the interests of expediency and efficiency, though, have at it. Anything that makes the controller's job easier will result in a better time of it for you.

Radio Strategies - the Controller's View

In the last chapter we offered two viewpoints on radio phraseology. But, how you word your transmissions is only part of the art of IFR communications. Regardless of your approach to the AIM guidelines, there are several things you can do to make the entire process work more smoothly.

In this chapter two controllers offer some specific guidelines on how to do it the right way...and some advice on what to avoid when you key the mic.

First up is Denny Cunningham, a tower controller at the proverbial "world's busiest airport," Chicago-O'Hare International. This section is devoted to how things work on the ground when things get insanely crowded—an everyday occurrence at O'Hare.

Lubricating the Communication Machine

Pilot: "Ground, Four Six Two's with ya."
Ground: "Four Six Two, say company and position."
Pilot: "Oh, sorry, that's TransOcean Four Six Two and we're off the right."

Gee, that narrows it down; I see Trans-Ocean airplanes clearing 14R, 9R, and 22R.

Ground: "TransOcean Four Six Two, what was the number of the runway you landed on, and where would you like to go?"
Pilot: "Oh, ah, we're off of 14 right, and we wanna go to the gate."

Take a deep breath. Chill out.

Ground: "TransOcean Four Six Two, could you be a little more specific? Your company has almost 50 gates on this airport."

Guess that sounded a little testy, but this guy is supposed to be a professional.

Pilot: "Oh, yeah, I guess so. We've got gate Kilo 24 at the international terminal, but it will be occupied for another 20 minutes. Guess we'll have to hold somewhere. Where would you like me to go?"

Boy, talk about a loaded question.

If there's one thing that can turn a busy controller into a belligerent one, it's a pilot who neglects the basic rules of air traffic communication. Tell the controller who you are, where you are, and what you want. To a harried controller, air time on the frequency has value, and he or she can't afford to squander it. And while it's safe to assume that a busy frequency equates to a busy controller, the opposite is not always true; a quiet frequency may mean that the paperwork and coordination that are part and parcel of a controller's job are about to overwhelm him.

Call for clearance

Let's start at clearance delivery, normally the first contact the IFR pilot will have on a flight from a controlled airport. The clearance delivery controller is probably working flight data, too, which is the paperwork position in the tower that processes clearances, makes ATIS recordings, and handles telephone coordination with adjacent facilities. At some locations, the same controller may be working ground control, too.

What this means is that much of the work the controller is doing isn't taking place on the clearance delivery frequency so your mental image of a fellow in a headset with a cup of coffee in one hand and the sports page in the other may not be quite accurate.

The best way to help the controller out is by first getting all the information you can from other sources. At locations where ATIS isn't available, listen in on the tower or ground frequency; you'll be able to pick up the basic information (wind, runway, and altimeter) in just a minute or two. When you call ground, be sure to use the phrase "have numbers" on initial contact.

If ATIS is available, of course, you'll start there, since it was specifically created to "...improve controller effectiveness and to relieve frequency congestion by automating the repetitive transmission of essential but routine information." ATIS also contains some not-so-routine information that might be of interest, such as gate-hold or other delay information, and temporary frequency changes ("clearance delivery frequency 121.6 is out of service; contact clearance on 119.25").

Remember to use the correct ATIS code when you contact the controller; to advise him that you "have numbers" when you have, in fact, listened to the entire ATIS broadcast makes it necessary for the controller to respond: "Verify you have received information alpha." If you haven't copied the information, the controller is required to provide it, unless you volunteer to obtain it. All of this means that getting the required information takes far more frequency time than it would have if they'd never invented ATIS in the first place.

Just the facts, please

Once you have the basic information in hand, contact clearance delivery. If you're more than 30 minutes before your proposed departure time, advise the controller of that on your initial call; he may have to make a special computer entry to get your flight progress strip. Other than that, though, the controller is just like Jack Webb on Dragnet—all he wants to hear are the facts: "O'Hare clearance, Twin Cessna One Two Three Four Quebec is at Butler Aviation, with echo, request IFR clearance to Dupage."

Have your pencil ready because as more facilities enter the computer age, the clearance will be sitting on a rack in front of the controller, who will read it to you immediately. At some locations, where the controller must make a telephone call to pick up the clearance (or, if the controller just needs some time to take care of a flaw in the clearance or some other duty), you may hear "clearance on request".

If that's the case, relax and wait for the callback. Constant cajoling just slows down the process as the controller takes time away from his telephones and computers to tell you he's working on it. If you're more than two hours past your proposed departure time, your clearance may have "timed out" and will no longer be in the system. You'll have to refile it.

In the event that you're not late, but your clearance still isn't in the computer, castigating clearance delivery is just shooting the messenger; the ball was most likely dropped at some prior point in the process, and regardless of whose fault it was, the clearance will have to be refiled. An accommodating controller may offer to do this for you, but be prepared to refile yourself if he's not so inclined. At some towers, controllers don't have the computer equipment necessary to enter a flight plan. For these reasons, it's good practice to obtain the clearance before starting engines.

Once you have your clearance secured, engine started, chocks pulled, and are completely ready to taxi, it's time to call the next controller in the chain. I've seen pilots call "ready to taxi" with a ground vehicle still attached to the nosewheel of a 400-ton airplane. Ground control: "Okay, clear to taxi. And will you be pushing that tug all the way to Denver?"

"Ready to taxi" should mean exactly that, and not that you hope to be ready in a few more seconds. The controller wants you to move *now.*

Ground metering

At the busiest of airports things can get nightmarishly congested, with the ramps and taxiways looking like rush hour in downtown Tokyo. Anywhere this can happen drastic measures are called for: hence the devising of "ground metering," a special controller position whose job is to relieve some of the frequency congestion on ground control by taking the initial calls for taxi, getting the appropriate paperwork together, and generally assisting the ground controller.

The folks who run O'Hare discovered long ago that the volume of traffic ruled out the use of procedures found at "normal" airports. This was especially true of ground control, where frequency congestion on ground control sometimes got so bad that the entire airport would grind to a halt until ground caught up.

Even before hubbing of airline flights, O'Hare traffic would ebb and peak in cyclic rushes. During a departure rush, the outbound ground controller would be inundated with traffic, all calling for taxi at once; when the "ready for taxi" calls overwhelmed the frequency, control of traffic already moving on the airport was impossible.

One response to this was to split ground control into two sectors, outbound (working departures taxiing out) and inbound (working arrivals from the runways into the gates). While this would appear to cut the ground controller's workload in half, it didn't really work out that way.

As traffic counts increased over the years, sheer volume began to overwhelm even this system. The end of one rush merged into the start of the next, causing traffic ready for taxi to compete for frequency time with aircraft calling for clearance. While delivery read the clearance and the pilot readback, another ten airplanes would be stalled in the alley between the concourses, ready to taxi but unable to get a word in on the frequency.

Meanwhile, like a bucket about to overflow, the arrivals kept pouring in. Arrivals at O'Hare can be landed on as many as five runways simultaneously; the key to survival for the inbound ground controller was (and still is) to keep traffic moving into the gates in a constant flow. Stop or even slow the flow for a minute, and you're in deep trouble.

That's because even a big airport like O'Hare has only so much real estate and it doesn't take that many airplanes to occupy every square inch of it. One or two wrong decisions by a ground controller—coupled with an overwhelmed clearance delivery—can grid-lock the entire airport.

In response to this, a new procedure required departures to contact clearance delivery twice—once to pick up the clearance, then again when ready to taxi. It was a good idea at the time. The rushes were well segregated, and once the clearances for a particular rush had been delivered, clearance got a breather and the controller working the position had time to field the "ready for taxi" calls and to generally assist outbound ground.

As traffic counts continued to increase, however, this scheme eventually proved unworkable and some fearsome backups resulted.

A particularly ugly example of this occurred when airplanes in the alleys couldn't move because of the congestion on clearance delivery. Since airplanes couldn't move out of the alleys, inbounds couldn't taxi to their gates. Where to put the inbounds? Parked nose-to-tail from the runways to the concourses. Sooner or later, the backlog reached critical mass and that meant a go-around for the next arrival, as the last was unable to clear the runway due to the gridlock. If things got bad enough, the entire system could back up, delaying departures at distant airports.

And so, the creation of ground metering, a position devoted exclusively to receiving "ready for taxi" calls, and expediting traffic movement out of the alleys to make way for the inbounds. The ground metering controller is an active assistant to the outbound ground controller, putting at the top of the list the airplanes that have to move first to avoid gridlock. This sometimes annoys pilots who called earlier, but whose aircraft are positioned in a such a manner that their immediate movement is not critical.

Sometimes ground has to issue "get 'em moving" taxi instructions that are the ground control equivalent of delaying vectors— they may not move the airplane in the direction he wants to go, but they keep the traffic moving in a flow away from the arrival runways.

During the past year, as the hub-and-spoke system expanded, airlines merged, and the "flow control" system proved unable to cope, the rushes have continued to increase in size and intensity. During the busy evening hours, outbound ground is often split yet again, so the ground metering controller now feeds two separate outbound grounds, each taxiing aircraft to several different runways.

Information on ground metering is usually contained in the ATIS broadcast, which is yet another reason to listen to ATIS first. If you call ground first, the controller won't know who you are since your flight progress strip—your ticket into the system—will be sitting at another position. The ground controller you've chosen to call may not even be the one you'll eventually talk to, since ground control chores are sometimes split up by function (inbound or outbound) or by other criteria.

If a ground metering position is in use, that controller will give you the proper frequencies to use, and generally, you'll be told to monitor the frequency. Although not yet defined in the AIM, the term monitor is becoming commonplace and has a specific meaning when used by ATC: Switch to the appropriate frequency and listen, but don't transmit—the controller knows you're there and will call you when your turn comes.

Your ready-to-taxi call, regardless of whom you make it to, should again include only the necessities: "O'Hare ground, Twin Cessna One Two Three Four Quebec is at Butler Aviation, with foxtrot, IFR, ready to taxi."

It's good practice to check the ATIS one last time before calling for taxi, and include the code for the latest ATIS you've received. And while the "IFR" comment may seem unnecessary, there will be times when the right hand doesn't know what the left is doing, and it's entirely likely that you'll taxi out and be airborne before anyone realizes that you think you're IFR, and they think you're VFR. Even though you have a ticket, it's best to make sure they haven't cancelled the show. Once you've heard your taxi instructions, acknowledge them in as succinct a manner as possible; at airports such as O'Hare, you may not get a chance to answer at all. The fact that you start moving will be acknowledgment enough. For routine instructions, this is fine, but if there are any sort of "hold short" restrictions imposed, a verbal acknowledgment is required.

If you're at all confused, particularly with regard to hold short instructions, tell the controller you're unfamiliar, or ask for progressive taxi instructions. Far better to suffer a slight delay that personal attention may require than to take a wrong turn into the middle of a landing runway.

Listen up on ground

Taxi to the departure runway may seem like a pretty cut-and-dried operation, but you'd be amazed how many problems are caused by pilots who are inattentive to ground control during taxi. Ground may call with new information about your IFR clearance or release, a change in routing to the runway, or a hold short instruction that he neglected to give you when you started your taxi.

You should monitor ground frequency in the run-up area as well, (unless instructed to change to tower) since that's the first place they'll look for you if something comes up.

Switch to the tower frequency when you're ready to go, but don't clog the frequency with useless information. During periods of light traffic, a quick "O'Hare tower, Twin Cessna One Two Three Four

Quebec is at 27 left, IFR, ready for takeoff," would be appropriate. But if the controller is up to his eyeballs in traffic, the "ready" call is a needless distraction.

Take a look at the traffic picture and try to guess what the controller's response to your call will be. If the answer is "roger," as it would be if you call "ready-in-sequence" with 6 airplanes ahead of you, hold your peace until it appears that you'll get a more positive response. The same applies when you're number one for the runway with traffic on final. Wait until it appears that the controller will say something other than "hold short."

What about those frustrating times when you're ready, the runway is clear, and you're still not allowed to launch? Well, as much as a good local controller likes to "move the metal," he's constrained by the rules and regulations of the controller handbook, as well as the wants and needs of the facility receiving his IFR departures. If a couple of good gaps on the final go by, or if the controller seems to be giving priority to traffic on another runway, don't become impatient. Chances are, the receiving facility has put you on "hold for release," and the tower controller is simply moving the traffic that is not so restricted.

Calling on arrival

Arriving at an airport on an IFR flight plan does have its advantages; the tower controller should know you're coming, and on what approach for which runway. But a simple "Three Four Quebec's with ya" won't make him happy. He may be predicating IFR separation between you and departure on your position report, so give him what he needs: "Dupage tower, Twin Cessna One Two Three Four Quebec, over ARNIE on the ILS runway 10, low approach, then VFR southwestbound."

If you'll be making a full-stop landing, the "who" and "where" will usually be sufficient, but if you're at an airport where practice approaches are common, adding the "what" (full stop, for instance) won't hurt. When your landing clearance includes instructions to hold short of another runway, acknowledge your intention to comply with these instructions; a laconic "roger" could come back to haunt both you and the controller if there's a misunderstanding.

If you don't feel comfortable landing in the distance available to the intersection, advise the controller immediately that you'll be unable to hold short. Phrases such as "we'll do our best" don't provide the assurance necessary to allow simultaneous intersecting runway operations. Let the controller know as soon as possible that another plan is necessary.

On the ground and slowing, pay attention to the tower controller's instructions. If there's a runway to cross between you and your destination on the field, facility policy may require that the tower controller keep you on his frequency until you cross that runway. Wait for the tower to tell you to contact ground before switching.

One thing many pilots fail to realize is that once you've turned off the active runway, you have effectively ceased to exist as an IFR airplane. At every point up until now, there have been handoffs from one controller to the next, so each controller pretty much knew who you were, where you were, and what you wanted before you talked to him.

Not so with ground control at your arrival airport. To him or her, you're a brand new customer, maybe an IFR arrival, maybe a VFR arrival, or maybe just an airplane requesting taxi from one point on the airport to another. Maybe you've landed here every day for the last 20 years and taxied to the Whiz-Bang Aviation tiedown area, but maybe the ground controller just got here yesterday, so he won't know that. Give him the information that he needs.

As we noted in the preface, once you've departed IFR your fate is in the hands of ATC. It's a good idea to keep the controllers happy, since they have a direct effect on how smoothly your flight will go.

Just flying by the book will get you through most situations with no problem, but there are times when expediency will get better results than a literal interpretation of the FARs.

Paul Berge, a tracon controller and instrument pilot, offers some advice for those who want to stay well clear of the wrath of an annoyed controller.

Ask Me Nicely

Every controller (and pilot) has some particular complaint that never seems to be corrected. The longer he flies or controls without the issue being resolved, the more this sore may fester, until one day it erupts on frequency in the form of some naive offender getting thoroughly trashed for being the one millionth violator.

This happened to me one day when I was working ground control. A 727 captain decided he'd had enough of our gate hold procedures. All the pent up bile from years of an uncorrected glitch spewed over me until I was forced to take the time-honored government employee's out: "Hey, I agree, but it's out of my hands; here, call my boss!"

Being a pilot, I know what sets pilots off. But you may not know what drives controllers nuts. So here's a sampling for your consideration. It's by no means complete.

Let's start with clearances. Reading back an entire approach clearance is always good way to irritate the hell out of a busy approach controller. "Ah roger, five from the marker; turn left heading 340; maintain 2500 until, etc., etc." Not a word missed. Also no turn, so the airplane blows through the localizer, forcing the controller to reclear the aircraft from the other side of the inbound.

Repeating a clearance is a good practice but don't forget to fly it. A successful localizer intercept is predicated upon the pilot turning the plane shortly after we issue the clearance. The margin for error is small, so clearances must be complied with promptly.

How should you acknowledge? How about this: "Three forty and 2500, cleared for the approach, Two Eight Delta."

Whether IMC or VMC, always acknowledge traffic calls. Nodding your head just doesn't work. Too often I've called traffic for a pilot only to get no response; call it again; no response. Finally, "Bonanza Two-Eight Delta, how do you hear?"

Answer (annoyed): "Loud and clear, and we're looking for that traffic." We need a response on frequency, even a useless "roger" is better than nothing.

Think before you talk. Keying the microphone while you're composing your response often produces nothing more than heavy breathing. (We sell these tapes to a 900 company in New Jersey.) Formulate your speech, then talk, then unkey. A surprising number of pilots keep the mike keyed (or it sticks) and we get treated to Hank and Mabel discussing dinner plans.

Speaking of stuck mikes, never, never sit on your mike in lieu of hanging it on the hook. I speak from experience. I once flew a Stits Skycoupe into Phoenix and after being cleared to land from many miles out, I placed the mike under my leg (more convenient, you know), and all the tower heard for five minutes was corduroy squeaking. Get a headset and a push-to-talk; saves cockpit work; saves face.

Similarly, don't turn down the comm volume to listen to passengers, instructors, Paul Harvey or that strange ticking noise that comes from the cowling whenever you enter the clouds. A volume knob turned down in IMC often remains down. "Gee, this frequency is awfully quiet lately." Flight instructors are often guilty of this. When things sound too quiet, pop the squelch to test audio continuity. That'll also detect a stuck mike.

Wanted: pireps

We need pireps; we don't need play-by-play reports of building weather. Generally, I solicit a pirep when I have time to copy one. They should be succinct: "Tops of the first layer 2000; bases of the next layer 2500; tops of everything 4000; clear above; no ice; smooth ride."

Not this, though: "Well, we went in about, oh, I'd say about twenty-five hundred feet MSL and then we didn't pick up any kind of ice to speak of; of course, I wasn't lookin' all that close, since we're solid right about now, except, hey, we seem to be breaking out...."

Somewhere below is a controller beating on the radar display waiting for the pirep to end, so a DC-9 can be given a vector toward the land of the living. If you're busy and can't give a pirep, say, "stand-by." Tell us later.

When you're in a busy terminal area being vectored for final, don't ask what your sequence is, unless you really need to know (e.g. minimum fuel; sick passenger; dinner reservations). Assume the controller wants to get you in as quickly as possible. Asking, "What's our number?" only slows things by tying up the frequency. You may have been number two before asking and find yourself bumped to number last. One thing that's guaranteed to provoke chaos in the radar room is a pilot who forgets to acknowledge a frequency change. We know you're busy, maybe terrified, but please, when the controller says, "Bonanza 28D, contact departure, good day," answer up. Say, "Bonanza 28D to departure." Or "Roger, 28D" or "Bye, bye." If you say nothing, we have to search for you.

"Anyone talking to that &%^!! 28D?"

Two clicks on the microphone barely constitutes an acknowledgement; we want to hear your voice (holds up in court better when FSDO is trying to take your license away).

If, at any time, you have a problem, special request or question, ask in a timely manner. Military pilots seem to be particularly prone to accepting a vector, reading it back, then announcing a low fuel status or request for flight break-up or a strafing run. Let us know before we take you out of your way.

And don't be afraid to question any clearance that doesn't sound right. I've cleared pilots to North Dakota who really filed to Florida. In a hurry, inside a dark tracon, I grabbed the wrong flight strip, one with a similar callsign.

Asking for practice approaches

A KC-135 pilot once informed me on frequency, "Approach! This is Major Bigstuff, and we burn umpteen pounds per hour. We've got three crews aboard here, all of whom need training, and we didn't fly all the way over here from O'Hare to be vectored around. We came here to shoot approaches! Now, what's the delay?"

I felt like telling him that the reason for the delay was that I never rose above the rank of E-5. It was Monday morning and I had a cold. And you, Major Bigstuff, have pushed all the wrong buttons. Of course, I didn't.

On the other hand, a Cessna 150 pilot calls with the appropriate dose of obsequiousness, requests several practice approaches, and, like magic, she's first—ahead of the kerosene-burning gorilla who thinks I might be intimidated by size, rank, or meanness of tone. (I am, actually, but only in face-to-face situations.)

The KC-135 pilot approached a delicate matter in a ham-fisted manner. According to the controller's handbook, 7110.65, practice approaches are never to delay other VFR or IFR traffic. Simply put, practice approaches are last. The Cheyenne pilot with a schedule to keep does not expect wide vectors to follow someone merely practicing.

So what's the best way to request practice approaches? Opinions vary, depending on parts of the country and complexity of airspace. Los Angeles approach might be a little reluctant to give a Cherokee multiple low approaches into LAX, whereas Casper, Wyoming usually has a neon sign blinking over the tower soliciting business.

Know your airspace. Get to know your ATC facility. Call the tracon and ask: "I'm new around here; what's the best way to get practice approaches?"

A trip to the local tracon on a slow evening can yield a wealth of goodwill and friendly tips. Call the facility during business hours and set up a tour of the facility. Keep a list of questions to ask about local airspace. The more pilots we can meet face-to-face, the more we can explain what works and what constitutes grounds for penalty vectors. Conversely, if you've had a problem with ATC, mention it tactfully.

One of the easiest ways to ensure access to practice approaches is to learn what times are best for such play. Every facility has slow times and rushes and these are usually predictable. Learn when the rushes happen and avoid them. At Des Moines, for example, Monday through Friday, our first departure flush begins at 7 a.m. and tapers off around 8:30 a.m. Another smaller rush hits at noon, after which things are quiet until about 4:30 p.m. when all the suits and ties come home for the night.

Thursday nights are double-discount nights. The local Air National Guard defends the skies against invasion from Minnesota—bad night to practice anything except noise complaint phone calls. Finally, weekends are up for grabs, with Sunday mornings deader than a beerless Shriner's convention. It's a good time to fly.

Consider the active runway before planning a training session. Why prepare a student for the ILS 12 approach, when all day the winds have been 300 at 20G40? Most likely, runway 30 is the active and approach probably doesn't want you grinding opposite direction to their departures. Go with the flow. Make as small a nuisance of yourself as possible and you're more likely to get what you want.

Try to picture what ATC has going on before jumping in with your request. Study the approach plate and picture where you think you'll fit into the traffic picture. If everyone else is being vectored to the final approach course for straight-in approaches, how would the controller respond to your request to shoot the full approach, procedure turn and all? Chances are there will be a pause after your request while the controller climbs back into his chair and says, "Hey, Fred, this guy thinks he can block all my traffic with a procedure turn! Let's say we spin him for awhile."

Let ATC know as soon as practical what you want. If you're filing a round-robin flight plan, add "IFR TRNG FLIGHT" in remarks or "REQUEST PRACTICE APPCHS." This tips off the controller that you plan to do more than land. If you have clearance delivery on the field, tell them what you want, "...have ATIS Echo; request IFR clearance for multiple practice ILS approaches at..." or "...request VFR practice ILS approach to full stop."

Listen to the ATIS well in advance. On initial callup tell the controller you have the information and, in the same breath, say, "requesting multiple practice ILS approaches." With that brief statement you've told the controller two things: exactly what you want (it may be different than what's advertised) and you've conveyed the impression that you understand the system and will be a pleasure to work with. Controllers sense weakness—sort of like wolves spotting a crippled lamb in the flock.

Take full advantage of satellite airports. If your home base has all the approaches you're used to, but only one runway is open, the ceiling is low, the controller sounds grumpy and you can hear 16 other pilots trying to get in, ask for an approach to an outlying airport.

Often satellite airport approaches go unused while everyone vies for a slot to home plate. Maybe you're not real good at this IFR stuff, anyhow, so what difference does it make where you make your mistakes?

Finally, be flexible. Don't try to outthink ATC (sometimes we can't even keep up with ourselves). Just get a general picture, a broad view if you can. Listen carefully before keying the mike; make your request and wait. You may hear things you've never imagined possible. Your plans may not jive with the traffic flow, and you may have to adjust, but, remember, real IFR may not go according to plan either.

Air filing and special requests

Problems often arise from air filed flight plans or flight plans filed on the ground but picked up airborne. The system is geared to pilots filing on the ground and being cleared from the ground up then back down. Skipping steps can cause confusion.

If you file a flight plan to be picked up enroute, make certain you know who to call for that clearance once aloft. Ask Flight Service and be prepared to be given the wrong frequency. They don't always understand what controllers do and may lead you astray.

On your initial callup, airborne, say, "Approach, Bonanza Two-Eight Delta, over PROPS intersection, 4500, with Echo (if you have it), IFR to DEN; I filed off Newton airport." Don't say "Ready to copy." We assume you're ready when you call.

Give the controller a few seconds to digest your callup. He may be searching through a stack of strips and will answer shortly. You may have flown into Center's airspace, and approach either doesn't have the flight plan or can't clear you because you are outside approach airspace.

Try to be aware of the vertical limits of approach airspace when calling for a clearance airborne. You may show yourself inside the lateral limits of their airspace but, unknowingly, be above it. Again, approach might tell you to contact the Center controlling the overlying airspace.

Be patient, flexible and stay VFR until cleared. Never say, "I'm in and out of these here clouds, so I need my clearance real soon." I've actually had pilots say words to that effect.

The more you know about the airspace through which you plan to fly, the less anxiety you'll encounter. If you're at an unfamiliar airport and preparing to file into untested skies, ask around. Ask local pilots how they fly a particular route. Call the ATC facility and ask us. "I want to file from here to there, is that possible?" If you can't get there from here, we'll tell you and make everyone's job easier and minimize the chance for error.

Mistakes happen and when they do, they should be addressed and corrected immediately. A pilot who errs certainly doesn't want the strong arm of the FAA coming down on him, so we see the occasional mode-C disappear, or radios become mysteriously silent while crew or controller decides a course of cover-up. No big deal. But little mistakes can lead to big problems. Changing frequencies without being told is more annoying than dangerous in most cases.

ATC annoyances

So far I've been speaking from the point of view of a controller. In all fairness, there are some things that controllers do that annoy me as a pilot, too.

No matter how good your radio work, you'll occasionally get spanked by a controller for missing a call or screwing up a readback. Not too surprisingly, this seems more common when there are two pilots in the cockpit. Single-pilot IFR requires a level of concentration and organization that seems to incorporate communications seamlessly into the overall scheme. Besides, when you're droning along in the clouds, there's not much else to do but listen to the radio.

Unless they're pilots, most controllers don't seem to understand that a two-man crew can get so busy for brief periods that whoever is doing the radio might be a little slow to answer a call or miss one entirely. And tape talks notwithstanding, some controllers have quirky habits that make them difficult to comprehend even when you can hear their words clearly. Every pilot has his or her own list of annoying things that controllers do. Here's mine:

The endless ATIS: Is it really necessary for an ATIS broadcast to be five minutes long? Basically, you need the ceiling and vis, winds, altimeter and runway in use, plus the alpha sequence. Personally, I don't need to know that the tango tiedown is muddy and that there are ditches on the airport access road. When you're in busy airspace—single-pilot or crew with one audio output—it's often necessary to listen to both approach and ATIS at the same time, a nerve wracking task. When the ATIS is overlong, it's that much harder to get the important stuff quickly.

Party time in the tower: Another complaint about ATIS. About once a month, I hear an ATIS broadcast with what sounds like the Mardi Gras going on the background. The clamor makes it impossible to hear the info, especially when performing the aforementioned two-radio trick. This happens on the approach frequencies now and again, too. While the controller is diligently vectoring away, his pals are yucking it up in the background, noise which eventually finds its way into the pilots' headsets, making clear communication all but hopeless.

Machine gun Mike: You can't blame a busy controller for talking fast. He or she's gotta move a lot of airplanes. But there are limits. When a clearance is delivered so quickly that even two pilots can't piece it together from the fragments they were able to copy, the entire process grinds to a halt when the controller is asked to repeat the entire thing, more slowly.

Related to this is the controller who paces everything so fast that it's nearly impossible to make a call without stepping on him or her. Sometimes it seems that a controller is actually trying to talk as fast as possible, regardless of whether it's really necessary or not.

Since I deal with the New York tracon almost exclusively, I've gotten used to the fast talkers. But when there's noise in the background—on any frequency or ATIS—I complain about it, either over the air or by telephone. Nothing hostile, just a polite request. Most of the time, the controllers aren't even aware of the problem and they're happy to fix it, either with a recut ATIS or shaddup to the noisemakers.

Get It Straight

In instrument flying, as in few other endeavors, miscommunication can be fatal. It is for this reason that pilots are supposed to repeat what controllers say to them. But, does this really ensure that the information has been passed between pilot and controller effectively?

Merely reading back a clearance without thinking about it has caused more than one accident. In this chapter O'Hare controller and Bonanza pilot Denny Cunningham explores the ins and outs of making sure that the instructions contained in a clearance get passed on properly, are understood and acted on accordingly.

Clear Clearance Readbacks

"Bonanza Zero Two Mike, three miles from WILLT, turn left heading 290, maintain 2100 until the localizer, cleared ILS 27L approach, 130 knots 'til WILLT, tower at WILLT, 120.75."

As I started the left turn, I keyed the push-to-talk and replied, "Roger, Bonanza Zero Two Mike."

The airline pilot/CFII in the right seat glanced in my direction, then shook his head. "I can't believe you did that."

"Did what?"

"Acknowledged a complicated clearance like that with 'roger.'"

Uh-oh.

Later, the discussion began. Basically, his position was that he reads back everything controllers say, both to insure that he has it right and to cover his hind end in the event that he got something wrong and the controller misses it on the readback. Better to hang together than hang separately, so to speak.

Naturally, since he is both an airline pilot *and* a CFII, I took this advice to heart and now give full and complete readbacks of every clearance I receive. Not!

Hear what you wanna hear

Okay, I know how important it is to be sure that what I think I heard is what the controller thinks he said. And I absolutely agree that if there is the slightest question in my mind as to exactly what was said, it would be foolhardy to not request immediate clarification.

I also know that we, as pilots, have a marked propensity for hearing what we expect to hear and responding accordingly. But what a lot of folks forget is that controllers are human too and they tend to do the same thing. If a controller tells a pilot to "Descend and maintain niner thousand," that's what he or she expects to hear in the readback. The pilot may be saying, "I'm inverted, on fire, and wish to report a midair collision with a red-suited guy in a sleigh..." and the controller won't bat an eye.

It's not unusual for a controller to not hear the readback at all, often due to blocked transmissions on busy frequencies, but sometimes because of his myriad off-air duties, which may range from coordinating with other facilities on the landlines to checking out the score of the basketball game on the boom box that's standard equipment in every ATC facility. The point is, the effectiveness of extensive readbacks is highly overrated.

I've seen a *lot* of operational errors (most of them as an interested observer, thank you) and almost none of them would have been prevented by the simple expedient of parroting everything the controller says. At the risk of being pilloried by the Conventional Wisdom crowd, I submit that long-winded readbacks are not a solution; they're part of the problem.

Why? Because the problem isn't usually one of a pilot misunderstanding a clearance that was meant for his airplane. The problem is more often a case of a pilot intercepting a clearance that was meant for somebody else. A full readback of the clearance doesn't do much to resolve such an identity crisis.

Whines and clips

Reasons for this abound, including sheer inattention on the part of any or all of the three parties involved (the controller, the pilot of the airplane intended to get the clearance, and the pilot of the airplane who intercepted it) and callsign "clipping." One major cause, however, is the sheer volume of traffic on the frequency (including the verbose readbacks), which often results in the callsign being blocked, sometimes during issuance of the initial clearance and sometimes during the readback.

In some cases, appropriate (meaning short) readbacks of the clearance are given, yet a preventable situation still turns to near catastrophe. A recent incident at O'Hare is typical:

The local controller was departing aircraft on three different runways: 32L at the T-1 intersection, 32R, and 9L. I've changed the names here to protect the guilty but use your imagination. Air Aardvark 73 was in takeoff position on 32R. Air Aardvark 873 was at runway 32L T-1 for departure (you can see where this is going). When the controller cleared Air Aardvark 873 for takeoff on 32L, Air Aardvark 73 departed 32R as well.

A careful listening to the tower tapes revealed that the controller involved had used impeccable phraseology, even including the runway number that the aircraft he was addressing was to depart on. The aircraft readbacks overlapped somewhat; what the controller heard was a slight squeal, followed by "...73, cleared for takeoff."

This is *almost* exactly what he would expect to hear. What made the situation somewhat ticklish was the fact that as Air Aardvark 73 went through the intersection of 9L, an MD80 that had departed that runway passed directly overhead, missing him by less than 200 feet.

So, what measures can a pilot take to prevent his being involved in such a scenario? In two words, situational awareness. Pay attention to *everything* on the frequency, not just transmissions that you believe to be for your aircraft.

Similar callsigns

Controllers do their best to advise the pilots of both aircraft if there are similar callsigns on the frequency, but even that method isn't fail-safe. In the case of Air Aardvark 73, there were at least two transmissions to Air Aardvark 873 (prior to the takeoff clearance and while Air Aardvark 73 was on the frequency) that might have alerted the pilot to the potential for confusion.

Air Aardvark 73's next clue that something was amiss should have been the fact that the takeoff clearance he heard included the words "...32L at T-1." Air Aardvark 73 was in position on 32R.

When a pilot is instructed to taxi into position and hold, controllers are required to advise a pilot of traffic departing a crossing runway. The pilot of Air Aardvark 73 received such an advisory. Approximately one minute before the controller issued the fateful clearance to Air Aardvark 873, the controller had cleared the MD80 for takeoff on 9L, again using the runway number as part of the clearance. In such a case, it's reasonable to assume that until the traffic on the intersecting runway clears the intersection, any takeoff clearance given was either for another aircraft altogether, or was issued in error.

An even more elementary solution was available to the pilot of Air Aardvark 73. He could've looked out the window. The day of the incident was good VFR. The 9L departure was in plain view from the approach end of runway 32R. My guess is that the sight of an MD80 bearing down on him at 140 knots would've made the pilot pause and reflect, so it's pretty obvious he didn't look.

Another question that every pilot should ask himself when he receives any clearance that he *thinks* is intended for his aircraft is, "Does this transmission make sense?" Controllers make mistakes. Using one callsign when they mean to use another is one of the more common. Correctly reading back a clearance that the controller intended for another guy may make you feel better when the time comes to play the tapes, but it can also put you closer to your Maker than you intended.

Pilot error, too

A number of years ago, the pilots of Delta 349 and Delta 439 found themselves nose-to-tail on the taxiway leading to their takeoff runway at O'Hare. When there was confusion as to which was first in line to cross another runway, both pilots took it in good humor as the controller attempted to determine that the airplanes and their paperwork were in the proper order. Just after Delta 349 was cleared for takeoff, with instructions to fly a 090 heading and expect a turn once airborne, Delta 439 was told to taxi into position and hold.

The next transmission by the controller made no sense at all. He told Delta 439, who was still in position on the runway, to "...turn right heading 180, and contact departure. Good day." The Delta flight, although not yet cleared for takeoff, dutifully sat on the runway and replied, "Right to 180, contact departure. Good day."

The controller, in a fit of verbal dyslexia, had transposed the callsigns. The readback, however, was exactly what he should have heard, so he didn't realize he'd made the error. Both pilots involved, oblivious to the fact that the previous instructions were clearly intended for an airborne airplane, did exactly what the controller told them to do.

Things got more exciting for both of them very shortly: the airborne airplane, still on the 090 heading instead of the 180 heading that the controller *thought* had been issued and acknowledged, swished a departure off another runway. The airplane in position on the runway very nearly had an arrival land on him due to the fact that he'd obediently switched to the departure control frequency and thus never heard the progressively more frantic takeoff clearances issued by the quite bewildered tower controller. Controller error? You bet. But he couldn't have done it without help.

Callsign last

So, just exactly what *should* a pilot read back? The number one most important item is the one most often neglected; the callsign, including company name or type aircraft, as appropriate. If there's an error, the only possible chance for correction is for the controller (or the other aircraft in the situation) to realize that the wrong guy just answered.

Simple "Rogers" or "We'll do all that" acknowledgements do nothing to help circumvent this most prevalent of errors. *Always* use the callsign. If the transmission wasn't for you, there's at least a reasonable chance that either the controller or the other pilot will and say so.

What's more, I prefer that the callsign be used not at the beginning of the clearance (as the AIM recommends) but at the end of the clearance, for two reasons: First, it's less likely to be the victim of an inadvertent clipping; two, if two airplanes reply to the same clearance, chances are good that although both pilots will *start* transmitting at the same time, they'll *end* at slightly different times, and the controller will probably be able to pick up at least one of the callsigns out of the clutter.

Second, it's always a good practice to give a short, concise readback of assigned headings, altitudes, or airspeeds. These are the key items that controllers use to keep one airplane from hitting another; any misunderstanding of these items could be critical. At least one air traffic facility, Chicago Center, has made readbacks of these items mandatory; if you don't read them back, the controller will ask you to.

Go beyond a simple readback if the transmission truly doesn't make sense in your present situation. Put the readback in the form of a question, if you think the instruction may have been issued in error. ("Center, this is Bonanza Zero Two Mike, did you want us to make a *right* turn to 250?") Be aware, of course, that the vagaries and inflexibility of the ATC system are such that a lot of things that don't seem to make sense to you ("We're going from Chicago to Seattle, and you want me to fly a 090 heading?") make perfect sense to controllers. Better safe than sorry, though.

Beyond these few items—callsign, headings, altitudes, airspeeds, and necessary clarifications—silence is golden. Frequency time is the workbench upon which a controller builds his product, which is safe and legal separation between airplanes. By all means, use it when you need to. But don't clutter it unnecessarily.

Index

H

L

P

R

S

V

W